Introduction to Lightwave Communication Systems

For a complete listing of the *Artech House Optoelectronics Library*, turn to the back of this book.

Introduction to Lightwave Communication Systems

Rajappa Papannareddy

Artech House
Boston • London

Library of Congress Cataloging-in-Publication Data
Papannareddy, Rajappa.
  Introduction to lightwave communication systems / Rajappa Papannareddy.
    p.   cm.
  Includes bibliographical references and index.
  ISBN 0-89006-572-1
  1. Optical communications.   2. Laser communication systems.
  3. Fiber optics.   I. Title.
  TK5103.59.P37   1997
  621.382'75—dc21   97-1921
                                                              CIP

British Library Cataloguing in Publication Data
Papannareddy, Rajappa
  Introduction to lightwave communication systems
  1. Optical communications 2. Optical fibres
  I. Title
  621.3'8275

  ISBN 0-89006-572-1

Cover design by Eileen Hoff

© 1997 ARTECH HOUSE, INC.
685 Canton Street
Norwood, MA 02062

International Standard Book Number: 0-89006-572-1
Library of Congress Catalog Card Number: 97-1921

10 9 8 7 6 5 4 3 2 1

Dedicated to my parents,
Papannareddy and Hanumakka

Contents

Preface

Information technology, a crucial component of the information superhighway, will have a major impact on how we communicate during the next several years. In recent years, lightwave communication systems, which are also known as optical fiber communication systems, have evolved from being the subject of basic research to commercial technologies. In fact, optical amplifiers have revolutionized the field of lightwave communication systems. Several textbooks on this subject have been written during the past few years, but many of these books have become outdated. Moreover, a balanced coverage of the lightwave technologies related to traditional and state-of-the-art devices and systems is lacking in many existing texts. Accordingly, this book aims to present a balanced coverage of the field of lightwave technologies with an emphasis on recent developments. Specifically, *Introduction to Lightwave Communication Systems*

- Provides material relating to various types of optical fibers, fiber nonlinear effects, light sources, photodetectors, passive components, optical amplifiers, optoelectronic integrated circuits, direct and coherent detection schemes, soliton systems, emerging networks, and multichannel systems;
- Presents practical aspects of the systems related to design and performance issues;
- Emphasizes a building-block approach to system design and analysis;
- Incorporates several field trials and experiments.

This book is mainly intended for senior-level undergraduate students specializing in the areas of electrical engineering, optics, or physics. Although it has been written primarily for senior-level undergraduate students, the book can also be used at a graduate level with an appropriate selection of topics. In addition, it serves as a good reference for practicing engineers and researchers. It is assumed that the background of the reader includes solid-state electronics and electromagnetics. In addition, a background relating to communication theory would be useful.

I have spent many weekends and nights in the preparation of the manuscript, and I would like to express my sincere thanks to my wife Manjula and my children

Prajwal and Priyanka for their patience and understanding. Finally, I thank Dr. Patricia Buckler of Purdue University North Central for reading the manuscript for grammatical errors and Mr. Dean Price for the wonderful artwork.

Rajappa Papannareddy
Valparaiso, Indiana
February 1997

Chapter 1

Introduction

Recent developments in lightwave communication systems have created many opportunities for telecommunications applications. Today, with the use of optical amplifiers, an optical fiber can carry higher data rates over greater distances in comparison with the other transmission media. The lightwave communication systems are now employed in many areas of telecommunications: intercity, interoffice, computer links, subscriber loop, and undersea systems. In this chapter, we describe the system evolution, basic concepts, advantages, and applications. Section 1.1 discusses the system evolution, and the basic concepts are discussed in Section 1.2. The system advantages are discussed in Section 1.3, and Section 1.4 describes the system applications.

1.1 SYSTEM EVOLUTION

A lightwave communication system in principle comprises a visible or near-infrared light source, whose carrier is modulated by the information signal; a fiber transmission medium; and a photodetector that detects the information signal. In the early 1960s, the demonstration of laser action [1,2] paved the way for optical communications. Nevertheless, it was the invention of a light source (light emitting diode (LED) [3] and a laser [4,5]) from a forward-biased gallium arsenide (GaAs) p-n junction that led to the feasibility of a lightwave transmission system. The laser is a coherent light source, that can launch more power into a fiber, can be modulated at higher speed, and has a narrower spectral width than the LED source.

Soon after, it was realized that some form of optical waveguide was essential to propagate the light. First, glass fibers having an attenuation of greater than 1,000 dB/km were considered, but the systems were impractical. Kao and Hockman at Standard Telecommunications Laboratories in England [6] studied the loss mechanisms in fibers and speculated the utilization of fibers for telecommunications applications. In 1970, Kapron and his colleagues at Corning glass produced a fiber with a loss of 20 dB/km [7]. The work toward reducing the fiber loss continued at several laboratories, and losses under 1 dB/km were obtained in pure silica fibers.

In addition to the low loss fibers, the photodetectors, such as positive-intrinsic-negative (p-i-n or PIN) diode and avalanche photodiode (APD), were developed to recover the information signal. The p-i-n diode yields a signal equivalent to one electron per incident photon, whereas an APD provides an internal amplification, yielding a signal equivalent to about 100 electrons per incident photon.

Following the successful demonstration of a low-loss optical fiber, research communities in the U.S., Europe, and Japan focused on the development of lightwave communication systems. The first systems were developed using multimode fibers (propagate many waveguide modes) in combination with an LED or laser source. Practical methods for splicing fibers and efficient coupling of source power into a fiber were found. As a first step in the transition from research to commercial applications, field trials carrying voice, data, and video signals were carried out in the 0.8 μm wavelength window. The bit rate-distance product (BL) was used as a figure of merit to measure the system capacity, where B is the bit-rate and L is the repeater spacing.

In the U.S., the first commercial system at 0.85 μm was deployed in the Chicago area during 1978. This system operated at a bit rate of 45 Mbps and achieved a repeater spacing of about 10 km [8]. Meanwhile, in England, Standard Telecommunications Laboratories carried out a field trial at 144 Mbps over a span of 9 km [9]. These systems utilized GaAs/AlGaAs semiconductor lasers/LEDs, multimode fibers, and direct detection receivers. The direct detection receiver recovers the information signal after converting the optical energy into electrical by using a photodetector and the associated receiver circuitry.

The course of further development was shifted to 1.3 μm wavelength window, because of the reduced fiber loss. Suitable sources and detectors were developed based on InGaAsP and Ge materials respectively for the 1.3 μm wavelength window. The practical implementation of these devices marked the beginning of the second generation of systems. Some of the earliest system designs used multimode fibers and LEDs instead of lasers because the LEDs offered low cost, simplicity of design, and high reliability for commercial applications. The systems operated at a bit rate up to 100 Mbps, with a repeater spacing of more than 20 km [10].

The big leap in performance of 1.3 μm systems came with the development of a singlemode fiber (propagates a single waveguide mode) which has both lower dispersion and lower loss than the multimode fibers. These systems belong to the third generation. In the early 1980s, the Eighth Trans-Atlantic (TAT-8) system was deployed under the Atlantic ocean spanning more than 5,600 km connecting the North American and European continents [11]. The recent systems in this window have attained a bit rate up to 10 Gbps, with a repeater spacing of over 80 km [12].

Next, to minimize the number of repeaters, the systems were designed around 1.5 μm wavelength region, to utilize the minimum loss (0.2 dB/km) of the single mode fiber, which resulted in a fourth generation of systems. But this wavelength region has the drawback of large fiber dispersion. To overcome this dispersion problem, several approaches were adopted. One of the approaches is to use a laser with extremely narrow spectral width, so called the single frequency laser (which oscillates in a single mode rather than in several modes). The successful development from this approach has resulted in a distributed feedback (DFB) laser [13]. In one experiment, transmission at 4 Gbps over a span of 103 km was achieved using a single mode fiber and a DFB laser [14].

In addition to the use of a DFB laser, the performances of fourth generation of systems were improved by shifting the minimum dispersion wavelength region of the fiber to the low loss 1.5 μm wavelength region, giving rise to a dispersion shifted fiber (DSF). Recent experiments have demonstrated the system operation up to 10 Gbps over a span of 80 km by using a dispersion shifted single mode fiber [15]. Also, a new method called dispersion-supported transmission (DST) has been utilized to overcome the fiber dispersion. Using this method [16], the systems have been operating at a bit rate up to 10 Gbps and have achieved a repeater spacing of 182 km.

During the mid 1980s, coherent detection techniques similar to radio systems were explored in optical fiber receiving systems [17,18]. The coherent detection techniques offer better receiver sensitivity and frequency selectivity than the direct detection schemes, and these systems belong to the fifth generation. The coherent detection schemes use an optical mixing scheme, called optical heterodyne or homodyne detection, which utilizes the coherent property of the laser. In the case of optical heterodyne detection, the resulting photocurrent from the mixing operation yields a signal at an intermediate frequency (IF), which carries the information signal. Whereas, in the case of optical homodyne detection, the information signal is directly detected through the optical mixing process, because the optical carrier wave is synchronous with the local oscillator wave. In 1984, the first coherent system experiments transmitted the signal at 100 Mbps over a span of 105 km using InGaAsP DFB lasers [19]. The recent systems have demonstrated the operation up to a bit rate of 4 Gbps with a repeater spacing of 202 km. Additionally, coherent schemes offer better channel capacity with the use of wavelength division multiplexing (WDM) and subcarrier multiplexing (SCM) techniques [20,21]. The typical characteristics pertaining to five generations of lightwave communication systems have been summarized in Table 1.1.

As lightwave communication systems evolved, computer data links such as local area networks (LANs), metropolitan area networks (MANs), and wide area networks (WANs) have replaced the coaxial cable or twisted copper wire systems with fiber optic cables. For intercity and interoffice systems, initially, it was the combination of low loss fiber and bandwidth that gave an economic advantage over coaxial systems. In the early 1980s, there began a steady growth of intercity fiber systems began in the United States, Europe, and Japan. In the United States, the first system FT3C multiplexed three 90 Mbps signals using wavelength division multiplexing scheme and carried about 240,000 voice circuits [22].

Table 1.1

Five Generations of Lightwave Communication Systems

Generation	System	Bit Rate (B)	Spacing (L)	BL Product (approximate)
I	0.85 μm, MMF	45–100 Mb/s	10 km	500 Mb/s-km
II	1.30 μm, MMF	100 Mb/s	20 km	2 Gb/s-km
III	1.30 μm, SMF	10 Gb/s	80 km	800 Gb/s-km
IV	1.55 μm, SMF	4 Gb/s	103 km	412 Gb/s-km
	1.55 μm, DSF	10 Gb/s	80 km	800 Gb/s-km
	1.55 μm, DST	10 Gb/s	182 km	1.82 Tb/s-km
V	1.55 μm, DSF	4 Gb/s	202 km	808 Gb's̄hs-km

Note: MMF: Multimode Fiber; SMF: Singlemode Fiber; DSF: Dispersion Shifted Fiber; DST: Dispersion Supported Transmission.

The advent of single mode fibers and the advanced optoelectronic components has transformed many of the megabit networks into multigigabit networks [23]. The evolution of present networks to the synchronous optical network (SONET) standard [24] (a synchronous frame structure for transmitting several time multiplexed voice channels) has created an infrastructure in the telecommunications industry. This infrastructure will support future broadband services such as video conferencing, high definition television distribution, and advanced interactive image communications. In recent years, hundreds of SONET systems operating at 2.488 Gbps have been installed using a single mode fiber with a repeater span of about 80 km. The SONET network at 10 Gbps is currently under installation, connecting Dallas and Los Angeles. Additionally, several new system concepts related to WDM local area networks, such as TeraNet, LAMDANET, RAINBOW, and STARNET, have emerged for high capacity applications [25] and these networks are described in Chapter 8. The telecommunications companies (telcos) have already begun the deployment of fibers in the loop, which is the region between the central office and the subscriber [26]. These subscriber loop systems are discussed in Chapter 8.

Beginning in the 1990s, the advent of an erbium-doped fiber amplifier [27] has revolutionized the field of lightwave transmission systems. In comparison with a regenerator, the amplifier device can boost the power of lightwave signals without the need for optoelectronic conversion and yields low cost. More importantly, the amplifier device is transparent to the data rate and the modulation format and can simultaneously accommodate many wavelength division multiplexed channels. Conversely, the system performance is limited by the accumulation of amplifier noise. In recent years, the terrestrial networks and undersea-cable systems that employ amplifiers have significantly increased their transmission capacity and operational flexibility. Systems based on direct detection schemes have achieved a total transmission distance of 9,000 km at 10 Gbps, utilizing 274 erbium-doped fiber amplifiers [28]. Similarly, the coherent detection schemes have achieved a transmission distance of more than 2,200 km at a bit rate of 2.5 Gbps utilizing twenty five in-line amplifiers, placed at approximately 80 km intervals [29].

The next most interesting innovation in lightwave communication systems is the soliton transmission, the optical pulses preserve their shape during propagation by compensating the fiber dispersion through nonlinearity. Although this method of transmission has been known for some years, it was only in the early 1990s, after the invention of an erbium-doped fiber amplifier, practical soliton transmission systems have been realized at many research laboratories. Recently, the fiber soliton experiment using optical time division demultiplexing (OTDM) technique has demonstrated a soliton data transmission of more than 12,200 km at a bit rate of 10 Gbps [30]. Furthermore, the system experiments up to 100 Gbps using a 50 km fiber have been carried out using OTDM techniques [31]. The soliton data transmission systems are discussed in Chapter 6.

The future lightwave communication systems are expected to operate at higher bit rates well beyond 10 Gbps and achieve longer repeater spacing through the use of optical amplification, wavelength division multiplexing, and soliton transmission techniques. Considerable advances in high-speed electronics, optoelectronic devices, system and network technologies are expected and these are the topics of today's research at many laboratories [32].

1.2 BASIC CONCEPTS

The main focus of this section is to describe the fundamental concepts of a lightwave communication system. Figure 1.1 shows the generalized block diagram of a long-haul lightwave link. The system consists of a transmitter, fiber transmission medium, receiver, in-line optical amplifiers, and associated couplers, connectors, and splices. This system in general can represent an analog or a digital link.

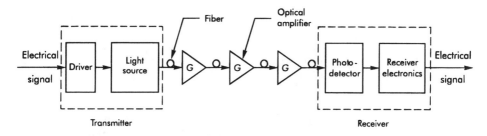

Figure 1.1 A generalized block diagram of a long-haul lightwave link.

1.2.1 Transmitter

The transmitter consists of a driver circuit along with a light source either a semiconductor laser or an LED. The light sources are described in Chapter 2. The modulation can be achieved in two ways: (1) by directly modulating the light source also known as intensity modulation or (2) by external modulation. In each case, the analog or digital modulation schemes can be employed. In the case of a digital modulation scheme, the first step involves the conversion of an analog signal into digital data ON/OFF (bit 1 or bit 0) pulses based on the pulse code modulation (PCM) technique. Then the ON/OFF pulses are line coded into different formats. In practical optical systems, line coding formats such as return-to-zero (RZ) or non-return-to-zero (NRZ) are used. These coding formats are shown in Figure 1.2. In the case of a RZ format, the pulse width is less than a full bit period, whereas the pulse width varies depending on the signal in the case of NRZ format. The NRZ line coding is often used in practice because of the smaller signal bandwidth. Next, the NRZ or RZ line coded signals are modulated to obtain a modulated optical carrier wave.

Depending on whether the amplitude or frequency or phase of the carrier wave is shifted between the ON/OFF pulses, several digital modulation schemes such as amplitude shift keying (ASK), frequency shift keying (FSK), and phase shift keying (PSK) can be utilized. These digital modulation schemes are depicted in Figure 1.3 for a NRZ line code format. The PSK scheme requires an external modulator whereas

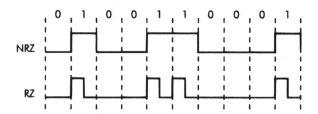

Figure 1.2 Return-to-zero and non-return-to-zero line coding formats.

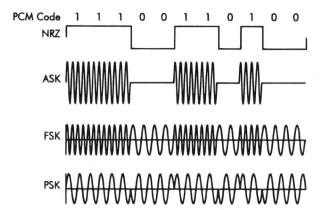

Figure 1.3 Basic digital modulation schemes: (a) ASK; (b) FSK; and (c) PSK.

the ASK and FSK schemes can utilize the direct or intensity modulation. Similarly, the analog modulation schemes such as amplitude modulation (AM) and frequency modulation (FM) can utilize the intensity modulation and whereas the phase modulation (PM) requires an external modulator.

The concept of intensity modulation can be explained with the use of laser diode characteristics assuming a digital signal input as shown in Figure 1.4. As can be seen from Figure 1.4, the laser output intensity is proportional to changes in the injected current, and hence it is called the intensity modulation. Also, note that the laser bias current in Figure 1.4 is less than the threshold value, which is recommended for digital signals. On the other hand, the laser bias current has to be higher than the threshold value for the case of analog signals for minimum signal distortion. In practice, the direct modulation may lead to changes in carrier density and causes frequency shifts in the carrier (chirp), which in turn leads to dispersion penalties. This problem can be overcome by the use of an external modulator, which is shown in Figure 1.5. In this scheme, the light generation and the modulation process are isolated, so that the chirp and turn-on transient effects are avoided. But the external modulators are polarization dependent, expensive and incur power loss. The external modulators are described in Chapter 7.

Another modulation technique is based on subcarrier analog modulation. In this case, the analog signal modulates a radio frequency subcarrier, which in turn modulates a light source as shown in Figure 1.6. In this scheme, amplitude or frequency modulation techniques can be employed. Additionally, pulse modulation schemes based on pulse position or pulse frequency modulation techniques can be

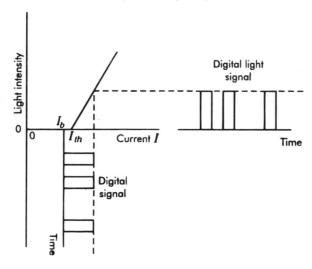

Figure 1.4 Intensity modulation for $I_b < I_{TH}$, where I_b is the bias current and I_{TH} denotes the laser threshold current. (*Source:* [33]. Reprinted with permission.)

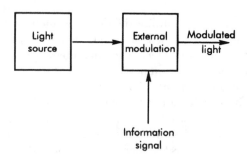

Figure 1.5 External modulation.

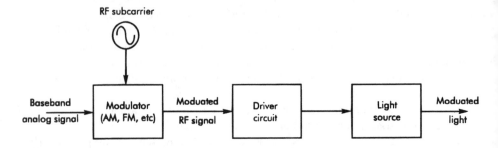

Figure 1.6 Subcarrier analog modulation.

utilized to transmit the analog information, which offer better signal-to-noise ratio in contrast with the intensity modulation.

1.2.2 Fiber Transmission Medium

The fiber transmission medium consists of a multimode or singlemode fiber for signal propagation. From the system point of view, the singlemode fiber is less dispersive, can carry higher data rates, and yields longer distance than the multimode fiber. On the other hand, the multimode fiber is less expensive and can be used for shorter distance communications. Typically the multimode fibers are used around 0.8 μm, whereas the single mode fibers are utilized at 1.3 and 1.55 μm wavelengths. When an optical signal propagates along a fiber, the power level is attenuated. The power level at a distance l km from the transmitter is [34]

$$P(l) = P_T 10^{-\alpha l/10} \tag{1.1}$$

where P_T is the power launched by the transmitter into the fiber and α is the attenuation constant of the fiber (dB/km). The fundamental processes that cause attenuation in fibers are Rayleigh scattering and absorption. The attenuation constant α is given by [34]

$$\alpha \approx \frac{0.6}{\lambda^4} \text{ dB/km} \tag{1.2}$$

where λ is the free-space wavelength in μm. The magnitude of the attenuation of the fiber versus wavelength is shown in Figure 1.7. From Figure 1.7, it can be seen that the minimum loss is near 1.55 μm and is about 0.2 dB/km. The maximum distance L over which a signal can be transmitted can be found from (1.1) as

$$L = \frac{10}{\alpha} \log_{10} \frac{P_T}{P_R} \tag{1.3}$$

where $P_R = P(l)$ represents the sensitivity of the receiver. It can be seen that from (1.3), the distance L mainly depends on the fiber attenuation and weakly depends on the transmitter and receiver power levels.

In addition to the fiber loss, different components of the signal travel at different speeds along a fiber giving rise to a pulse spread, which is commonly known as fiber modal dispersion. The dispersion in the fiber limits the bit rate of the signal as well as the transmission distance. The modal fiber dispersion occurs mainly in multimode fibers. In the case of single mode fibers, the primary mechanism is material dispersion, which results from the frequency dependence of the refractive index of the fiber material. The material dispersion can be further controlled by reducing the spectral width of the optical source or by using a dispersion shifted fiber. The bandwidth of

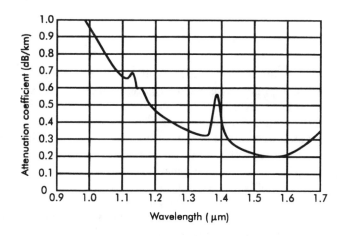

Figure 1.7 Fiber attenuation curve. (*Source:* [35]. Reprinted with the permission of McGraw-Hill.)

the fiber at each wavelength is mainly limited by the fiber dispersion. For each wavelength, we use the *BL* product as a general indication of the practical capability of the system. The *BL* product in general depends on the type of fiber, amount of dispersion, and spectral purity of the light sources. The fiber transmission media and the fiber dispersion effects are discussed in Chapter 2.

1.2.3 Receiver

At the receiving end of an optical link, the receiver consists of a photodetector either a PIN or an APD device along with the appropriate receiver electronics. The photodetector converts the optical signal into an electrical signal. The photodetectors are discussed in Chapter 2. The lightwave receiving systems can be classified into direct and coherent detection schemes. In the case of a direct detection scheme, the information signal is detected after converting the photon energy into electrical. Conversely, the signal is detected by mixing the lightwave signal with the receiver local oscillator in the case of coherent detection. However, from the system point of view, the coherent receivers add complexity into the system. The direct and coherent receivers are described in Chapters 4 and 5 respectively.

The performance of a receiver is degraded by factors such as laser linewidth, relative intensity noise of the source, receiver noise, fiber nonlinear effects, and polarization mismatch. These effects will have an impact on the maximum transmission distance and repeater spacing. In the case of a digital lightwave link, the performance of a receiver is primarily measured by the bit-error-rate (BER). Conversely, in an analog lightwave link, the performance of the receiver is measured by the signal-to-noise-ratio (SNR) or the carrier-to-noise-ratio (CNR).

1.2.4 Optical Amplifiers

The system links usually utilize optical amplifiers, placed at regular intervals along the fiber, (in-line amplifiers) instead of conventional repeaters to compensate for the losses in a link. Additionally, an amplifier can be used as a booster for the transmitter or as a front end amplifier for the receiver. The optical amplifiers are classified into semiconductor amplifiers and fiber amplifiers. In comparison with semiconductor amplifiers, fiber amplifiers offer high gains, low splice losses and reflections, low intermodulation distortion, and are polarization insensitive. In recent years, the erbium-doped fiber amplifiers have proven to be efficient and reliable for long-haul communications. In practical long-haul systems, the accumulation of amplifier noise, fiber dispersion as well as the spectral purity of the laser sources impose a limit on the maximum number of in-line amplifiers. The optical amplifiers are described in Chapter 3.

1.2.5 Optoelectronic Integrated Circuits

The operating speed of lightwave communication systems has increased dramatically over the last several years. Currently, research is focused on 20- and 40 Gbps systems. As the operating speed increases, the monolithically integrated circuits (ICs) play an important role in the practical realization and commercialization of high speed systems. The ICs based on field effect transistor (FET) and heterojunction bipolar transistor (HBT) technologies are rapidly maturing [36] and can offer higher operating speed in the near future. Additionally, the silicon/bipolar technology has been utilized to develop IC's for the transmitter and receiver modules up to 10 Gbps [37], and ICs for over 20 Gbps are under development [38]. The optoelectronic ICs are discussed in Chapter 7.

1.3 SYSTEM ADVANTAGES

The lightwave communication systems offer several advantages in comparison with the other guided wave transmission systems. In this section, we will summarize several advantages of lightwave communication systems.

1.3.1 Low Transmission Loss

As compared with the coaxial cables, or copper wires, the optical fibers have very low loss yielding longer spacing between the repeaters. For example, the repeater spacing for the lightwave systems can be greater than that for microwave systems by one or two orders of magnitude. Additionally, the amplifiers can be utilized to obtain larger spacing.

1.3.2 Wide Bandwidth

In comparison with any other guided media, the optical fibers offer the largest bandwidth possible for a variety of system applications. The available bandwidth for the single mode fibers in the 1.5 to 1.7 μm transmission window is approximately 24,000 GHz.

1.3.3 Nonelectromagnetic Interference

The optical fiber medium is immune to the radio frequency interference problems, ground loop problems, and lightning-induced interference.

1.3.4 Small Size and Weight

The size of optical fiber cable is very small in comparison with conventional wiring (twisted copper pair or coaxial). Additionally, the weight of the fiber optic cable is much less than that of a conventional wire. This is very advantageous for telcos, because most telephone cabling is put through underground ducts, which often are crowded or full. Hence, the capacity can be increased by using fiber optic cables.

1.3.5 Small Crosstalk

The fiber optic cable is not a conductive medium, and hence it is not susceptible to the neighbor signals or to external signals that results in a very small crosstalk. This is a useful characteristic, because it eliminates the extra cost of shielding the cable.

1.3.6 Privacy and Security

As optical fibers do not radiate energy, their signals cannot be detected by others. This offers a degree of privacy and security.

1.3.7 Cost

The cost is a very important factor in the system design. The cost comparison should include the costs of installation, operation, and maintenance with an equivalent competing technology such as twisted copper pair or coaxial. The other competing systems such as digital microwave radio, cellular systems, and satellite links use free transmission media. Traditionally, the cost of lightwave systems are compared with the twisted copper pair or coaxial cable. For long-distance communications, single mode fiber appears to have significant economic advantage over twisted pair copper and coaxial cables, particularly where large bandwidth and loop lengths are involved. Alternatively, multimode fibers are economically attractive for LANs and other computer data links.

1.4 SYSTEM APPLICATIONS

The lightwave system encompasses the transmission of voice, video, and data signals to a variety of applications. In this section, we briefly discuss several applications of lightwave communication systems.

1.4.1 Voice

The voice transmission includes telephone trunk, interoffice, intercity, subscriber loops, and transoceanic applications. The telephone trunk system is based on different digital signaling levels. For example, at level 1, the carrier has an information carrying capacity of 1.54 Mbps which corresponds to 24 multiplexed voice channels. It is important to note that many channels can be transmitted over one fiber, whereas with a conventional wiring system, the voice channels transmitted on a digital carrier would have to be split among many individual pairs of wire. Furthermore, in recent years, the SONET based fiber optic networks have increased the system operating rate up to 10 Gbps.

The voice transmission is not only limited to the terrestrial links but also covers the transocean links. The undersea lightwave transmission systems TAT-8 to TAT-13 link the east coast of U.S. with Europe. The TAT-9 to TAT-13 systems use 1.55 μm wavelength and operate at a bit rate much higher than 1 Gbps. Other undersea lightwave systems such as TPC-3 and HAW-4 have been installed in the Pacific Ocean to link Hawaii, Japan, Guam, and California.

1.4.2 Video

Video is used for television industry, remote monitoring, and surveillance applications. The broadcast television industry uses lightwave transmission for short distances such as studio to the transmitter or live event to the equipment van. The cable television industry picks up signals from several sources, including antennas, satellite, and microwave links. All the signals from these sources are transmitted simultaneously along a single fiber with the use of frequency division multiplexing and a radio subcarrier. Additionally, the lightwave systems can be used to transmit video signals for surveillance and remote monitoring.

1.4.3 Data

The data applications include computer networks such as interoffice data links, LANs, MANs, and WANs. The interoffice data links include the connection of different computers with different buildings within a university campus or a city. The LAN connects various computers within a building or a region. The MAN is used to link computers within the metropolitan area, and the WAN is used for linking various cities in the same region. The fiber optic networks are discussed in Chapter 8.

At present, the cable TV companies and telcos are competing with each other to combine audio, video, and data signals on a fiber to provide broadband services related to TV broadcast channels, video-on-demand, interactive TV, and computer data links. Other applications of fiber transmission include avionics, satellite Earth stations, submarines, military systems related to command, control, and communication links on ships and aircrafts.

Problems

Problem 1.1

Assume a typical lightwave communication system that has a bit rate distance product of 10 Gbps/km.

(a) Calculate the distance to transmit 100 Mbps, 200 Mbps, and 500 Mbps of data.
(b) What wavelength windows should be used for each of the bit rates in (a)?

Problem 1.2

Find the attenuation constant of a fiber at 0.8, 1.3, and 1.55 μm.

Problem 1.3

Find the transmission distance of a fiber having an attenuation constant of 0.2 dB/km, given $P_R/P_T = 0.1$.

Problem 1.4

A lightwave communication link has a length of 10 km and has a splice loss of 1.0 dB. Assume a 0.85 μm system with a fiber loss of 3 dB/km and connector losses of 1.0 dB. The system link requires a receiver sensitivity of −56 dBm for proper operation. Calculate the required optical power level of the transmitter.

Problem 1.5

Repeat the above problem with an ideal in-line amplifier having a gain of 20 dB.

References

[1] Schawlow, A. L., and Townes, C. H., Infrared and optical masers, *Phys. Rev.*, Vol. 112, 1958, pp. 1940–49.
[2] Maimon, T. H., "Stimulated optical radiation in ruby," *Nature*, Col. 187, 1960, pp. 493–494.
[3] Keyes, R. J., and Quist, R. M., "Recombination radiation emitted by gallium-arsenide," *Proc. of I. R. E.*, Vol. 50, 1962, pp. 1822–1823.
[4] Hall, R. N., Fenner, G. E., Kingsley, J. D. Soltys, T. J., and Carlson, R. O., "Coherent light emission from *GaAs* junctions," *Apply. Phys., Lett.*, Vol. 9, p. 366, 1962.
[5] Nathan, M. I., Dumke, W. P., Burns, G., Dill, F. H., Jr, and Lasher, G., "Stimulated emission of radiation from GaAs p-n junctions," *Apply. Phys., Lett.*, Vol. 1, p. 62, 1962.
[6] Kao, C. K., and G. A. Hockman, "Dielectric-fiber surface waveguides for optical frequencies," *Proc. IEE*, Vol. 113, 1966, pp. 1151–1158.
[7] Kapron, F. P., D. B. Keck, and R. D. Maurer, "Radiation losses in glass optical waveguides," *Apply. Phys. Lett.*, Vol. 17, 1970, pp. 423–425.
[8] Cannon, T. C., D. L. Pope, and D. D. Sell, "Installation and performance of the Chicago lightwave transmission system," *IEEE Trans. Commun.*, Vol. COM-26, 1978, pp. 1056–1060.
[9] Hill, D. R., Jessop, A., and Howard, P. J., "A 144 Mbits/s field demonstration system," *Conf. Proc. Eur. Conf. Opt. Commun.*, 1977, p. 240.

[10] Gloge, D., A. Albanese, C. A. Burrus, E. L. Chinnock, J. A. Copeland, A. G. Dentai, T. P. Lee, T. Li, and K. Ogawa, "High speed digital lightwave communications using LED's and PIN Photodiodes at 1.3 μm," *Bell Sys. Tech. J.*, Vol. 59, 1980, p. 1365.

[11] Runge, P. K., and P. R. Trischitta, "The SL Undersea lightwave system," *Journal of Lightwave Technology*, Vol. LT-2, 1984, pp. 744–753.

[12] Miyamoto, Y., K. Hagimoto, K. Nakagawa, T. Kagawa, H. Tsunetsugu, M. Ohhata, and N. Tsuzuki, "10 Gb/s 1310 nm optimized fiber link over 80 km using laser-diode transmitters and avalanche-photodiode receivers with high manufacturability," *Tech. Dig., OFC/IOOC '93* , San Jose, CA, 1993, Paper TUD2, pp. 13–15.

[13] Matsuoka, T, H. Nagai, Y. Itaya, Y. Noguchi, U. Suzuki, and T. Ikegami, "CW operation of DFB-BH *GaInAsP/InP* lasers in 1.5 μm wavelength region," *Electronics Letters*, Vol. 18, 1982, pp. 27–28.

[14] Gnauck, A. H., B. L. Kasper, R. A. Linke et al., "4 Gb/s transmission over 103 km of optical fiber using a novel electronic multiplexer/demultiplexer," *IEEE Journal of Lightwave Technology*, Vol. LT-3, 1985, pp. 1032–1035.

[15] Fujita, S., N. Henmi, I. Takano, "A 10 Gb/s, 80 km optical fiber transmission experiment using a directly modulated DFB-LD and a high-speed InGaAs-APD," *Tech. Dig., OFC '88*, New Orleans, 1988, PD-16.

[16] Wedding, B., Franz, B., and Junginger, B., "10 Gb/s optical transmission up to 253 km via single-mode fiber using the method of dispersion supported transmission, *IEEE Journal of Lightwave Technology*, Vol. 12, No. 10, 1994, pp. 1706–1719.

[17] Saito, S., Y. Yamamoto, and T. Kimura, "Optical heterodyne detection of directly frequency modulated semiconductor signals," *Electronics Letters*, Vol. 16, 1980, pp. 826–827.

[18] Okoshi, T., "Recent progress in heterodyne/coherent optical-fiber communications," *IEEE Journal of Lightwave Technology*, Vol. LT-2, No. 4, 1984, pp. 341–345.

[19] Emura, K., Shikada, M., Fujita, S., Mito, I., and Minemura, K., "Novel optical FSK heterodyne single filter detection system using a directly modulated DFB-laser diode," *Electronics Letters*, Vol. 20, No. 24, 1984, pp. 1022–1023.

[20] Glance, B., et al., "Densely Spaced WDM Coherent optical star network," *Electronics Letters*, Vol. 23, No. 17, 1987, pp. 875–876.

[21] Hill, P. M., and R. Olshansky, "Multigigabit subcarrier multiplexed coherent lightwave system," *IEEE Journal of Lightwave Technology*, Vol. 10, no. 11, 1992, pp. 1656–1664.

[22] Stauffer, J. R., "A lightwave system for metropolitan and intercity applications," *IEEE Journal on Selected Areas in Communications*, Vol. SAC-1, No. 3, 1983, p. 413–419.

[23] Rolland, C., L. E. Tarof, and A. Somani, "Multigigabit networks: The challenge," *IEEE Mag. of Lightwave Commun. Sys.*, Vol. 3, No. 2, 1992, pp. 16–27.

[24] Ballart, R., and Y. C. Ching, "SONET: Now it's the standard optical network," *IEEE Commun. Mag.*, Vol. 27, No. 3, 1989, pp. 8.

[25] Kazovsky L. G., C. Barry, M. Hickey, C. A. Naronha, and P. Poggiolini, "WDM local area networks," *IEEE Mag. Lightwave Commun. Sys.*, Vol. 3, No. 2, 1992, pp. 8–15.

[26] Burpee, D., ed., "Special issue on fiber in the local loop," *IEEE Mag. of Lightwave Commun. Sys.*, Vol. 1, No. 3, 1990, pp. 6–30.

[27] Mears, R. J., L. Reekie, I. M. Jauncey, and D. N. Payne "Low noise erbium-doped fibre amplifier operating at 1.54 μm," *Electronics Letters*, Vol. 23, 1987, pp. 1026–1028.

[28] Taga, H., et al., "10 Gb/s 9000 km IM-DD transmission experiments with 274 Er-doped fiber amplifier repeaters," *Tech. Dig. OFC/IOOC '93*, San Jose, CA, 1993, paper PD1.

[29] Saito, S., T. Imai, and T. Ito, "An over 2200 km coherent transmission experiment at 2.5 Gb/s using erbium-doped-fiber in-line amplifiers," *IEEE Journal of Lightwave Technology*, Vol. 9, no. 2, 1991, pp. 161–169.

[30] Suzuki, M., N. Edagawa, H. Taga, H. Tanaka, S. Yamamoto, and S. Akiba, "10 Gb/s, over 12200 km soliton data transmission with alternating-amplitude solitons," *IEEE Photonics Technology Letters*, Vol. 6, no. 6, 1994, pp. 757–759.

[31] Kawanishi, S., H. Takara, K. Uchiyama, T. Kitoh, and M. Saruwatari, "100 Gb/s, 50 km optical transmission employing all-optical multi/demultiplexing and PLL timing extraction," *Tech. Dig., OFC/IOOC '93*, San Jose, CA, 1993, Paper PD2.

[32] Heidemann, R., B. Wedding, and G. Veith, "10-GB/s transmission and beyond," *Proc. of IEEE*, Vol. 81, no. 11, 1993, pp. 1558–1567.

[33] Kashima, N. *Optical Transmission for the Subscriber Loop*, Artech House, Norwood, 1993.

[34] Henry, P. S., "Lightwave Primer," *IEEE Journal of Quantum Electronics*, Vol. QE-21, No. 12, 1985, pp. 1862–1877.

[35] Winch, R. G., *Telecommunication Transmission Systems*, McGraw-Hill, New-York, 1992.

[36] Cowles, J., A. L. Gutierrez-Aitken, P. Bhattacharya, and G. I. Haddad, "7.1 GHz bandwidth monolithically integrated *InGaAs/InAlAs* PIN-HBT transimpedance photoreceiver," *IEEE Photonics Technology Letters*, Vol. 6, No. 8, 1994, pp. 963–965.

[37] Miyamoto, Y., K. Hagimoto, M. Ohhata, T. Kagawa, N. Tsuzuki, H. Tsunetsugu, and I. Nishi, "10 Gb/s strained MQW DFB-LD transmitter module and superlattice APD receiver module using *GaAs* MESFET IC's," *IEEE Journal of Lightwave Technology*, Vol. 12, No. 2, 1994, pp. 332–342.

[38] Ichino, H., M. Togashi, M. Ohhata, Y. Imai, N. Ishihara, and E. Sano, "Over 10 Gb/s IC's for future lightwave communications," *IEEE Journal of Lightwave Technology*, Vol. 12, No. 2, 1994, pp. 308–319.

Chapter 2

System Description

An overview of a lightwave communication system was presented in Chapter 1. This chapter is devoted to the optoelectronic devices and components involved in the generation, transmission, and processing of lightwave signals. In recent years, these devices and components have been discussed extensively in many textbooks and review articles. The objective of this chapter is to review the basic characteristics of these devices and components from the stand point of system design. Optical fibers are discussed in Section 2.1, and light sources are discussed in Section 2.2. Section 2.3 describes the photodetectors and Section 2.4 discusses the optical fiber cables, splices, and connectors. The source-fiber coupling is discussed in Section 2.5. The driver circuits for sources are presented in Section 2.6, and finally, the system design issues are discussed in Section 2.7.

2.1 OPTICAL FIBERS

An optical fiber is a cylindrical dielectric waveguide that guides the light in a direction parallel to its axis. In 1966, Kao and Hockman [1] studied the wave propagation in glass optical fibers for signal transmission which eventually led to the field of lightwave communication systems. When the low loss optical fiber was realized in 1970 [2], significant progress was made toward understanding the properties of the fiber as an information transmission medium. During the past twenty five years, many excellent books [3–13] and review articles [14–18] on lightwave communications have been written.

The optical fibers are made of glass consisting of silica or a silicate (SiO_2). Sand is the raw material for silica. The optical fiber has two regions: (1) the inner core region and (2) the outer cladding region. These two regions have two different refractive indices. The cladding region must have a lower refractive index than the inner core region. The oxide dopants such as GeO_2, B_2O_3, and P_2O_5 are added to silica to obtain the refractive index difference. The fibers with glass cores and glass claddings (called all-glass fibers) yield low loss and hence are suitable for long-distance applications. For short-distance applications (less than 100 meters), plastic clad fibers (PCD) can be utilized, because these are less expensive and have higher mechanical strength than the all-glass fibers. The PCD fibers are made of silica core with cladding of a plastic material such as polymer or teflon, but these fibers exhibit high loss. Also, all-plastic fibers are viable for short distance communications and they exhibit higher losses and are durable. The core as well as the cladding regions use plastic materials in the case of all-plastic fibers.

The popular methods used to fabricate glass fiber are (1) vapor-phase axial deposition (VAD) and (2) modified chemical vapor deposition (MCVD). For a more detailed treatment of fiber materials and fabrication methods, the reader is referred to the literature [19].

The basic concepts of optical fibers were introduced in Chapter 1. In this chapter, we study the important optical fiber structures and their properties associated with the transmission characteristics. Section 2.1.1 discusses the basic structures of optical fibers, and the electromagnetic wave propagation is discussed in Section 2.1.2. The fiber bandwidth and dispersion aspects are discussed in Section 2.1.3. The nonlinear effects in optical fibers and their effects on system performance are described in Section 2.1.4.

2.1.1 Optical Fiber Structures

The transmission characteristics of an optical fiber is mainly governed by its structural properties which in turn establishes the information carrying capacity. The optical fiber structures are classified based on the refractive index profiles (step index or graded index) or based on the core diameter (single mode or multimode) relative to the wavelength. A single-mode or multimode fiber can be manufactured with either step or graded index profiles. However, in practice, single-mode fibers use step index profile and multimode fibers use graded index profile to reduce modal dispersion. Furthermore, the multimode fibers can be classified into different groups based on the numerical apertures, and core and cladding diameters [13]. In the following sections, we discuss the fiber structures, and their properties associated with the transmission characteristics.

2.1.1.1 Step Index Fiber

Figure 2.1 shows the cross-sectional view of a step index fiber along with refractive index profile and signal path. The inner region of the cylinder is called the core of the fiber having a refractive index n_1. The core is surrounded by a solid dielectric cladding region with a refractive index n_2. It can be seen that the refractive index is uniform throughout the core, but it changes at the core/cladding interface. The propagation of light along a fiber can be explained by means of electromagnetic waves called the modes. Each propagation mode consists of electric and magnetic

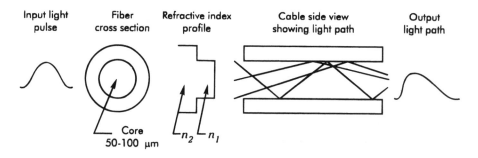

Figure 2.1 Structure, refractive index, and signal transmission in a step index fiber.

field lines. The step-index structure supports many modes and hence it is known as a multimode fiber.

The rays that propagate in a fiber are (1) meridional rays and (2) skew rays. The meridional rays are confined to the meridian planes of the core. These rays can be classified into two types: (1) bounded rays that are confined to the core and (2) unbounded rays that are refracted out of the core as per the laws of geometrical optics. On the other hand, the skew rays are not confined to any plane and generally are more difficult to track as they travel along the fiber. To understand the basic principle of lightwave propagation, we use the ray trajectory analysis based on the meridional rays as shown in Figure 2.2. The ray trajectory analysis is a valid approximation provided the radius of the core is much larger than the signal wavelength; otherwise, electromagnetic analysis is required.

Consider a ray of light entering the fiber core at an incident angle θ_A as shown in Figure 2.2. The ray travels from a medium of higher refractive index n_1 (core region) into a medium of lower refractive index n_2 (cladding region). If the ray strikes the core-cladding interface at a certain angle (called the critical angle), then from Snell's law, the ray would suffer total internal reflections. This ray follows zig-zag path along the fiber which is the basic mechanism for light confinement in fibers or

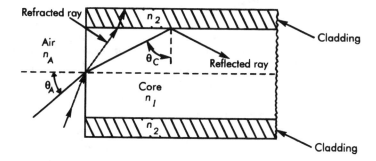

Figure 2.2 Ray trajectory in a step-index fiber.

if the ray strikes the interface at angles less than the critical angle, then the ray will suffer refraction. From Snell's law, the critical angle θ_C required for total internal reflections is given by

$$\sin \theta_C = \frac{n_2}{n_1} \tag{2.1}$$

where n_1 and n_2 are the refractive indices of the core and cladding regions respectively. Also, by applying Snell's law at the air-core interface, it can be shown that the maximum acceptance angle for total internal reflections is given by

$$n_A \sin \theta_M = n_1 \sin \theta_C = \sqrt{(n_1^2 - n_2^2)} \tag{2.2}$$

where n_A denotes the refractive index of the air, θ_M is the maximum acceptance angle. Equation (2.2) defines an important fiber parameter called the numerical aperture (NA) as

$$NA = n_A \sin \theta_M = n_1 \sin \theta_C = \sqrt{(n_1^2 - n_2^2)} = n_1\sqrt{2\Delta} \tag{2.3}$$

Where

$$\Delta = \frac{(n_1^2 - n_2^2)}{2n_1^2} \cong \frac{(n_1 - n_2)}{n_1}$$

is the relative core-cladding difference which is typically $\leq 2\%$. NA for typical telecommunication fibers varies from 0.1 to 0.2. The plastic fibers have an NA greater than 0.5.

2.1.1.2 Graded Index Fiber

If an optical signal is injected into a multimode fiber, the signal energy will be distributed among the large number of modes. Each of these modes travel with its characteristic group velocity along the length of the fiber. In the case of a step index fiber, the transit time difference between the fastest and the slowest propagation modes is higher and hence causes pulse spreading (called intermodal dispersion). The intermodal dispersion which affects the information capacity of the fiber will be discussed in Section 2.1.3. However, if the refractive index profile of the core varies with the radial distance r from the center of the fiber, then the transit time difference between the modes can be made small, which improves the information capacity of the fiber. The most popular form of refractive index profile is given by

$$n(r) = n_1\left[1 - 2\Delta\left(\frac{r}{a}\right)^s\right] \quad r < a \qquad (2.4a)$$

$$n(r) = n_1(1 - 2\Delta)^{1/2} \quad r \geq a \qquad (2.4b)$$

where n_1 is the refractive index at the core axis, r is the radial distance from the core axis, a is the core radius, n_2 is the refractive index of the cladding, s defines the shape of the profile, and n_2 is the relative index difference. The parameter s can be varied to get different profiles. For example, if $s = 1$, (2.4a) yields a triangular profile, $s = 2$ a parabolic, and $s = \infty$ simplifies to a step index profile. Equations (2.4a,b) define the graded index fiber. The graded index fiber's structure with a parabolic index profile is shown in Figure 2.3.

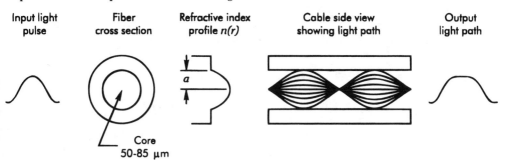

Figure 2.3 Cross-sectional view of a graded-index fiber along with the refractive index profile and signal path.

2.1.1.3 Single Mode Fiber

The single mode fiber structure is shown in Figure 2.4. Its fabrication is similar to that of a step-index fiber structure, but the difference in core size and dopant levels between the two types results in different operating characteristics.

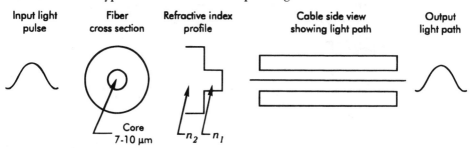

Figure 2.4 Structure, refractive index, and signal transmission in a single mode fiber.

The earlier structures of a SMF are optimized for the 1.3 μm wavelength window for minimum dispersion which are commonly known as conventional single mode fibers (C-SMF). These structures use step index profiles with a small core size of typically 8–12 μm with cladding size of 125 μm. New structures have been developed to combat the effect of dispersion at 1.55 μm wavelength window. These are known as dispersion shifted single mode fibers (DS-SMF) and dispersion flattened single mode fibers (DF-SMF). In the case of DS-SMF, the zero dispersion wavelength of single mode fibers can be shifted to the low-loss 1.55 μm wavelength region by compensating the material dispersion with increased waveguide dispersion. This can be accomplished through the reduction of the core size as well as by increasing the relative index difference [20–23]. Some of the popular design structures are shown in Figure 2.5.

Conversely, the dispersion flattened structures yield a flat dispersion over the low-loss region between 1.3 and 1.6 μm window [21,22]. These structures use multi-layer index profiles and typical refractive index profiles are shown in Figure 2.6.

2.1.2 Wave Propagation in Fibers

In Section 2.1.1, we studied the light confinement in a fiber using a ray trajectory approach. This approach is valid only for a moderate value of Δ and the fiber core diameter should be large compared to the signal wavelength. To get an accurate representation for wave propagation in fibers, electromagnetic analysis based on wave equations is essential. First, we review Maxwell's equations and then derive a wave equation. For a homogenous medium, the Maxwell equations take the form

(a)   (b)

Figure 2.5 Typical refractive index profiles for dispersion shifted fibers: (a) the triangular-index core and (b) the segmented triangular-index core.

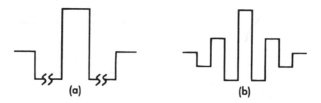

(a)   (b)

Figure 2.6 Typical refractive index profiles for dispersion flattened fibers: (a) depressed-index clad and (b) segmented core dispersion flattened.

$$\nabla \times E = -\mu \frac{\partial H}{\partial t} \tag{2.5}$$

$$\nabla \times H = -\epsilon \frac{\partial E}{\partial t} \tag{2.6}$$

where E and H represent the electric and magnetic filed intensities respectively, the parameter μ represents the permeability of the medium and ϵ is the dielectric constant. In the above equations, E and H are the vector fields. Now, the electromagnetic wave equation can be derived from Maxwell's equations. Taking the curl of (2.5) and then using (2.6), we get

$$\nabla \times (\nabla \times E) = -\mu \frac{\partial}{\partial t} (\nabla \times H) = -\epsilon\mu \frac{\partial^2 E}{\partial t^2} \tag{2.7}$$

But $\nabla \times (\nabla \times E) = \nabla(\nabla \cdot E) - \nabla^2 E = \nabla^2 E$, because $\nabla \cdot E = 0$. Then (2.7) becomes

$$\nabla^2 E = \epsilon\mu \frac{\partial^2 E}{\partial t^2} \tag{2.8}$$

which is the wave equation in terms of electric magnetic fields. Similarly, one can obtain the wave equation in terms of magnetic fields by taking the curl of (2.6) and following the above procedure, we get

$$\nabla^2 H = \epsilon\mu \frac{\partial^2 H}{\partial t^2} \tag{2.9}$$

Next, let us consider the electromagnetic wave propagation in a cylindrical fiber as shown in Figure 2.7.

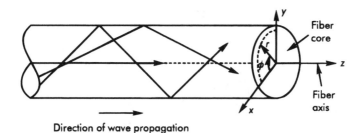

Direction of wave propagation

Figure 2.7 Electromagnetic wave propagation in a cylindrical fiber. (*Source:* [12]. Reprinted with the permission of McGraw-Hill.)

In Figure 2.7, r denotes the radial distance of the fiber, ϕ denotes the angular distribution of the wave, and z is the direction of wave propagation. The electromagnetic fields in the cylindrical coordinate system can be represented by

$$E = E_0(r, \phi)e^{j(\omega t - \beta z)} \tag{2.10}$$

$$H = H_0(r, \phi)e^{j(\omega t - \beta z)} \tag{2.11}$$

Note that the fields given by (2.10) and (2.11) are dependent on time t and the direction of propagation z. The parameter β is called the propagation constant, which represents the z component of the propagation vector. Substituting (2.10) and (2.11) into (2.5) and (2.6), we get the r, ϕ, and z components of the E and H field vectors as

$$\frac{1}{r}\left(\frac{\partial E_z}{\partial \phi} + j\beta r E_\phi\right) = -j\mu\omega H_r \tag{2.12a}$$

$$j\beta E_r + \frac{\partial E_z}{\partial r} = j\mu\omega H_\phi \tag{2.12b}$$

$$\frac{1}{r}\left(\frac{\partial (rE_\phi)}{\partial r} - \frac{\partial E_r}{\partial \phi}\right) = -j\mu\omega H_z \tag{2.12c}$$

$$\frac{1}{r}\left(\frac{\partial H_z}{\partial \phi} + j\beta r H_\phi\right) = j\epsilon\omega E_r \tag{2.12d}$$

$$j\beta H_r + \frac{\partial H_z}{\partial r} = -j\epsilon\omega E_\phi \tag{2.12e}$$

$$\frac{1}{r}\left(\frac{\partial (rH_\phi)}{\partial r} - \frac{\partial H_r}{\partial \phi}\right) = j\epsilon\omega E_z \tag{2.12f}$$

If E_z and H_z are known, then the transverse components E_r, E_ϕ, H_r, H_ϕ can be determined in terms of E_z and H_z, by using the separation of variables. The resulting equations are

$$E_r = -\frac{j}{p^2}\left(\beta\,\frac{\partial E_z}{\partial r} + \frac{\omega\mu}{r}\,\frac{\partial H_z}{\partial\phi}\right) \tag{2.13a}$$

$$E_\phi = -\frac{j}{p^2}\left(\frac{\beta}{r}\,\frac{\partial E_z}{\partial\phi} - \omega\mu\,\frac{\partial H_z}{\partial r}\right) \tag{2.13b}$$

$$H_r = -\frac{j}{p^2}\left(\beta\,\frac{\partial H_z}{\partial r} - \frac{\omega\epsilon}{r}\,\frac{\partial E_z}{\partial\phi}\right) \tag{2.13c}$$

$$H_\phi = -\frac{j}{p^2}\left(\frac{\beta}{r}\,\frac{\partial H_z}{\partial\phi} + \omega\epsilon\,\frac{\partial E_z}{\partial r}\right) \tag{2.13d}$$

where

$$p^2 = \omega^2\mu\epsilon - \beta^2 = k_0^2 n^2(r) - \beta^2$$

Using (2.13c) and (2.13d) in (2.12f), we get the wave equation for a fiber in cylindrical system of coordinates as

$$\frac{\partial^2 E_Z}{\partial r^2} + \frac{1}{r}\frac{\partial E_Z}{\partial r} + \frac{1}{r^2}\frac{\partial^2 E_Z}{\partial\Phi^2} + [k_0^2 n^2(r) - \beta^2]E_Z = 0 \tag{2.14}$$

where E_Z represents the z-component of the E(electric) field,

$$k_0 = \omega\sqrt{\epsilon_0\mu_0} = \frac{2\pi}{\lambda}$$

is the free space wave number, $n(r)$ is the refractive index profile of the fiber, and ω is the angular frequency of the wave. Similarly, using (2.13a) and (2.13b) in (2.12c), we get a wave equation in terms of the magnetic field as

$$\frac{\partial^2 H_Z}{\partial r^2} + \frac{1}{r}\frac{\partial H_Z}{\partial r} + \frac{1}{r^2}\frac{\partial^2 H_Z}{\partial\Phi^2} + [k_0^2 n^2(r) - \beta^2]H_Z = 0 \tag{2.15}$$

The waves or modes which satisfy the above wave equations subject to the boundary conditions are indeed the solutions to the above equations. If $E_Z = 0$ results in a transverse electric (TE) modes. Similarly, if $H_Z = 0$, then the modes are called transverse magnetic (TM) modes. On the other hand, if both the fields are non zero, then the modes are known as hybrid (HE or EH) modes which normally exist in a fiber. Equation (2.14) or (2.15) can be solved for waves propagating in the z-direction, using the separation of variables method, which assumes a solution of the form

$$E_Z \propto E_z(r)e^{jm\phi}e^{j(\omega t - \beta z)} \tag{2.16}$$

where the functional dependence on ϕ is represented by the function $e^{jm\phi}$, where m can be a positive or a negative integer. The other components E_r, E_ϕ, H_r, H_ϕ, can be found in terms of E_Z or H_Z by using (2.13a)–(2.13d). Substituting (2.16) into (2.14) gives

$$\frac{\partial^2 E_z(r)}{\partial r^2} + \frac{1}{r}\frac{\partial E_z(r)}{\partial r} + \left[k_0^2 n^2(r) - \beta^2 - \frac{m^2}{r^2}\right]E_z(r) = 0 \tag{2.17}$$

which is a well-known differential equation for Bessel functions. In the following paragraphs, we use (2.17) to solve for the propagation modes in step index and graded-index fibers.

2.1.2.1 Modes in a Step Index Fiber

For a step index fiber, the solutions for the guided modes must remain finite for the region inside the core as r becomes very small ($r \to 0$) and the solutions must decay for the region outside the core as r becomes very large ($r \to \infty$). Then (2.17) can be written as two differential equations for the two regions.

$$\frac{\partial^2 E_z(r)}{\partial r^2} + \frac{1}{r}\frac{\partial E_z(r)}{\partial r} + \left[k_0^2 n_1^2 - \beta^2 - \frac{m^2}{r^2}\right]E_z(r) = 0 \qquad (r < a) \tag{2.18}$$

$$\frac{\partial^2 E_z(r)}{\partial r^2} + \frac{1}{r}\frac{\partial E_z(r)}{\partial r} - \left[k_0^2 n_2^2 - \beta^2 + \frac{m^2}{r^2}\right]E_z(r) = 0 \qquad (r > a) \tag{2.19}$$

The solutions to (2.18) and (2.19) are given by

$$E_z = A'J_m(ur)e^{jm\phi}e^{j(\omega t - \beta z)} \qquad r < a \tag{2.20}$$

$$E_z = C'J_m(wr)e^{jm\phi}e^{j(\omega t - \beta z)} \qquad r > a \tag{2.21}$$

Similarly, one can set up the identical equations for the H_z term. In this case, the solutions for H_z term inside and outside the core are given by

$$H_Z = B' K_m(ur) e^{jm\phi} e^{j(\omega t - \beta z)} \qquad r < a \tag{2.22}$$

$$H_Z = D' K_m(wr) e^{jm\phi} e^{j(\omega t - \beta z)} \qquad r > a \tag{2.23}$$

where A', B', C', and D' are arbitrary constants, and

$$u = \sqrt{(k_0^2 n_1^2 - \beta^2)}$$

and

$$w = \sqrt{(\beta^2 - k_0^2 n_2^2)}.$$

The terms $J_m(ur)$ is the Bessel function of the first kind and order m and $K_m(wr)$ is the modified Bessel function of the first kind and order m. It is known that $K_m(wr) \rightarrow e^{-wr}$ as $wr \rightarrow \infty$. Also, $K_m(wr) \rightarrow 0$ as $r \rightarrow \infty$ for $w > 0$, which implies that $\beta > k_0 n_2$, representing the cutoff condition for the region outside the core (a mode is no longer bound to the core region). For the region inside the core, another condition on β results from $J_m(ur)$. In this case the parameter u must be real for $E_z(r)$ to be real, which implies that $k_0 n_1 > \beta$. Therefore, for a mode to be guided, the propagation constant β must satisfy the following condition

$$n_2 k_0 \leq \beta \leq n_1 k_0 \tag{2.24}$$

The value of β can be found from the boundary conditions. The boundary conditions require that the tangential components (both E and H fields) must be the same at the core-cladding interface ($r = a$). First let us consider the tangential components of E_z

$$E_{z1} - E_{z2} = A' J_m(ua) - C' K_m(wa) = 0 \tag{2.25}$$

where E_{z1} and E_{z2} are the z components of the electric field at the inner core–cladding interface and at the outer core–cladding interface respectively. Similarly, the tangential components of the H field are given by

$$H_{z1} - H_{z2} = B' J_m(ua) - D' K_m(wa) = 0 \tag{2.26}$$

The ϕ components of the E field can be found from (2.13b). $E_{\phi1}$ at $r < a$ can be found by substituting (2.20) and (2.22) into (2.13b). Similarly, $E_{\phi2}$ at $r > a$ can be found by substituting (2.21) and (2.23) into (2.13b). The resulting $E_{\phi1} - E_{\phi2}$ equals [12]

$$-\frac{j}{u^2}\left[A'\frac{jm\beta}{a}J_m(ua) - B'\mu\omega uJ'_m(ua)\right]$$

$$\frac{-j}{w^2}\left[C'\frac{jm\beta}{a}K_m(wa) - D'\mu\omega wK'_m(wa)\right] = 0 \qquad (2.27)$$

Similarly, for the ϕ components of the H field, the resulting $H_{\phi 1} - H_{\phi 2}$ equals

$$-\frac{j}{u^2}\left[B'\frac{jm\beta}{a}J_m(ua) + A'\epsilon_1\omega uJ'_m(ua)\right]$$

$$-\frac{j}{w^2}\left[D'\frac{jm\beta}{a}K_m(wa) + C'\epsilon_2\omega K'_m(wa)\right] = 0 \qquad (2.28)$$

Using (2.25)–(2.28), one can set up a matrix in terms of the coefficients and the constants. From the determinant of this matrix, the eigenvalue equation for β can be written as [12]

$$(I_m + L_m)(K_0^2 n_1^2 I_m + k_0^2 n_2^2 L_m) = \left(\frac{m\beta}{a}\right)^2\left(\frac{1}{u^2} + \frac{1}{w^2}\right)^2 \qquad (2.29)$$

where $I_m = J'_m(ua)/uJ_m(ua)$ and $L_m = K'_m(wa)/wK_m(wa)$. Equation (2.29) can be solved for β for the permissible range given in (2.24).

The transcendental (2.29) is generally solved by using numerical methods. In general, there will be n roots of (2.29) for a given m value. The corresponding roots are designated by β_{mn} and the modes are denoted by TE_{mn}, TM_{mn}, EH_{mn} or HE_{mn}. Where the first integer m denotes the azimuthal mode number, signifies the azimuthal variation of the field and the second integer n is the radial mode number which specifies the radial variation of the field. When $m = 0$ in (2.29), then the right-hand side of (2.29) vanishes, resulting in TE_{0n} or TM_{0n} modes.

When $m \neq 0$, then the exact solution to (2.29) requires numerical methods. However, if the core and cladding indices are nearly the same ($\Delta << 1$) then an approximate solution to the transcendental equation would result. This approximation leads to weakly guided modes. In this case, all the fiber modes are linearly polarized. Hence HE_{mn} modes are assumed to degenerate and are commonly called LP_{mn} (linearly polarized modes). In this case, the resulting transcendental equation is [24]

$$\frac{uJ_{m-1}(ua)}{J_m(ua)} = -w\frac{K_{m-1}(wa)}{K_m(wa)} \qquad (2.30)$$

The above equation can be solved for discrete propagation constants for different guided modes. Next, let us understand the different modes resulting from (2.30). In general, several modes can exist in a fiber. A mode is said to be cut off, if the wave does not propagate. To find the cutoff conditions, let us define a parameter, called the normalized frequency or V-number as

$$V = \sqrt{u^2 + w^2} = ak_0\sqrt{n_1^2 - n_2^2} \cong ak_0n_1\sqrt{2\Delta} \qquad (2.31)$$

V is a dimensionless number which determines the number of modes in a fiber. The V-number can be represented in terms of the normalized propagation constant b as

$$b = \frac{a^2w^2}{V^2} = \frac{(\beta^2/k_0^2 - n_2^2)}{n_1^2 - n_2^2} \Rightarrow \beta \approx k_0n_2(1 + b\Delta) \qquad (2.32)$$

From (2.32), it can be inferred that for a guided mode $0 \le b \le 1$ and the modes are cut-off when $\beta/k_0 = n_2$. In terms of the normalized parameters b and V, (2.30) can be written as

$$V\sqrt{1-b}\,\frac{J_{m-1}(V\sqrt{1-b})}{J_m(V\sqrt{1-b})} = -V\sqrt{b}\,\frac{K_{m-1}(V\sqrt{b})}{K_m(V\sqrt{b})} \qquad m \ge 1 \qquad (2.33a)$$

$$V\sqrt{1-b}\,\frac{J_1(V\sqrt{1-b})}{J_0(V\sqrt{1-b})} = V\sqrt{b}\,\frac{K_1(V\sqrt{b})}{K_0(V\sqrt{b})} \qquad m = 0 \qquad (2.33b)$$

A plot of b versus V is shown in Figure 2.8, and it can be seen that each mode exists only for a certain value of V.

Let us consider some examples to understand the mode propagation in a fiber. As a first example, let us examine the curve for $V = 2.4048$ (corresponding value of b is 0.53). In this case, only a single mode HE_{11} (designated as LP_{01}) exists in a fiber is called the fundamental mode, which is the principle of a single mode fiber. Equation (2.33b) suggests that the first root of $J_0(V)$ and $J_1(V)$ occurs when $V = 2.4048$. This condition can be used to find the cutoff wavelength of the fiber as [24]

$$\lambda_C = \frac{2\pi an_1\sqrt{2\Delta}}{2.4048} \qquad (2.34)$$

This equation can be used to design a single mode fiber for a given core size a and Δ. Also, from (2.34), it can be inferred that the cutoff wavelength of HE_{11} mode is zero only if the core radius is zero. All other modes except HE_{11} have cutoff wavelengths above which no wave propagation occurs. To understand (2.34), consider a fiber for 1.3 μm wavelength, with $n_1 = 1.48$ and $\Delta = 0.01$, the required core radius from equation (2.34) is about 2.4 μm. Decreasing the value of Δ, will increase the core

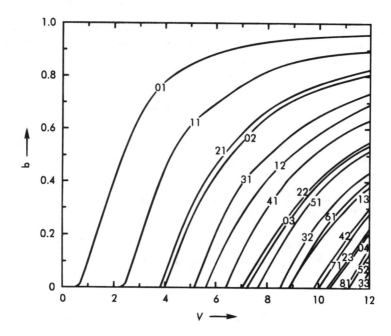

Figure 2.8 Normalized propagation constant b vs V for various LP_{lm} modes. Here $LP_{lm} \leftrightarrow HE_{(l+1)m}$, $EH_{(l-1)m}$ for $l \neq 1$ and $LP_{1m} \leftrightarrow HE_{2m}$, TE_{0m}, TM_{0m} for $l = 1$. (*Source:* [5]. Reprinted with the permission of Academic Press.)

radius. The core radius for single mode fibers is typically in the range of 3–5 μm. As a second example, consider when $V \geq 2.4048$. Specifically, when $2.4048 \leq V \leq 3.8317$, then modes such as TE_{01}, TM_{01}, HE_{21} (designated as LP_{11}) will result, which is the case of multimode fibers.

Other important design parameters of single mode fibers are: (1) the mode field profile and (2) the mode field diameter. The mode field profile gives the profile of the intensity distribution of the mode, which is approximately Gaussian, whereas the mode field diameter represents the width of the mode field profile. Gaussian approximation methods have been used to obtain the modal characteristics of single mode fibers [25,26].

2.1.2.2 Modes in a Graded Index Fiber

The exact analytical solution to (2.17) can be obtained if the refractive index $n(r)$ is uniform in a fiber. However, in the case of a graded-index fiber, an approximation method based on WKB technique (named after Wenzel, Kramers, and Brillouin) is commonly employed to analyze the modal characteristics of the graded index fiber. The applicability of this method requires the following condition to be satisfied.

$$\left| \frac{1}{n} \frac{dn}{dr} \right| \geq \frac{1}{\lambda} \qquad (2.35)$$

which implies that the refractive index is a slowly varying function of the spatial coordinate. The above condition is generally satisfied in the case of a multimode graded-index fibers. The first step in applying WKB technique is to eliminate the term dE_z/dr by letting $U = r^{1/2}E_z(r)$ in (2.17), which yields [12]

$$\frac{d^2U}{dr^2} + k^2(r)U = 0 \qquad (2.36)$$

where $k^2(r) = k_0n^2(r) - \beta^2 - (m^2 - 1/4)/r^2$. For a guided wave in a fiber, one must have

$$\int_{r_1}^{r_2} k(r)\, dr \approx \int_{r_1}^{r_2} \left[k_0^2 n^2(r) - \beta^2 - \frac{m^2}{r^2} \right]^{1/2} dr \approx (2m - 1)\frac{\pi}{2} \approx m\pi \qquad (2.37)$$

where $l = 1, 2, \ldots$ is an integer, which represents the radial mode number. Equation (2.37) represents the eigen value equation under WKB approximation for a graded-index fiber. The integration limits for (2.37) are governed by the solution of $k(r) = 0$, such that $k(r)$ is real for $r_1 \leq r \leq r_2$, which implies that the field is oscillatory within this region and $k(r)$ is imaginary outside that region, leading to exponentially decaying fields. The solution of (2.37) will yield the propagation modal constant β_{mn} for different combinations of m and n. Using equation (2.37), it can be shown that the maximum number of guided modes in a graded-index fiber is given by [12]

$$N_{\max} = \frac{s}{s + 2}\, a^2 k_0^2 n_1^2 \Delta \qquad (2.38)$$

Equation (2.38) can be approximated to $V^2/2$ for a step index fiber since $s = \infty$ and for a parabolic refractive index profile ($s = 2$) the number of guided modes is approximated to $V^2/4$. Thus for a given V, the total number of guided modes in a step index fiber is twice that of an equivalent (both having the same a and Δ) parabolic profile fiber.

2.1.3 Fiber Bandwidth and Dispersion

As we studied in Chapter 1, the transmission bandwidth of the fiber and the fiber dispersion are inter-related. The fiber dispersion causes the pulse broadening, and hence these pulses overlap as they propagate along the fiber and incur data errors at the receiver. From the system point of view, it is important to understand the mechanism of dispersion in fibers as well as the effect of dispersion on the fiber bandwidth. The information capacity of the system is measured in terms of the fiber bandwidth or the *BL* product. We choose *BL* product to characterize the information capacity of the system. First, We will examine the effect of pulse spread on the fiber bandwidth then the mechanism of dispersion in fibers is discussed.

2.1.3.1 Fiber Bandwidth

Consider a linear time invariant system which is shown in Figure 2.9. The optical fiber medium is assumed to follow the laws of the linear time invariant system. However, the fiber medium can be modeled with a complex transfer function which yields an approximate Gaussian impulse response for the fiber [4]. However, for simplicity, let us assume a narrow optical pulse at the fiber input. Using convolution it can be shown that the corresponding response at the fiber output is $h(t)$. The transfer function of the fiber is then given by

$$H(\omega) = \int_{-\infty}^{\infty} h(t)e^{j\omega t}\, dt \tag{2.39}$$

Assuming a symmetric $h(t)$ and upon the expansion of $h(t)$ by using Taylor's series, the transfer function of the fiber can be approximated as [27]

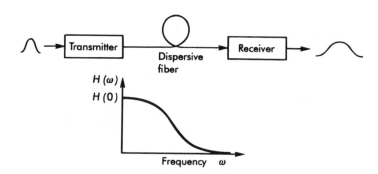

Figure 2.9 Effect of pulse broadening on the frequency response of the fiber. (*Source:* [27]. © 1985 IEEE. Reprinted with permission.)

$$H(\omega) \cong H(0)\left(1 - \frac{\omega^2\sigma_p^2}{2}\right)$$ (2.40)

where σ_p^2 represents the mean square width of $h(t)$ which is also known as the mean square impulse spread. Note that as the frequency ω increases, the frequency response of the fiber $H(\omega)$ decreases. This effect is quite significant when data sequence such as $101010 \cdots$ (alternate 1's and 0's have the highest frequency content) are transmitted. In this case, most of the power is contained in a Fourier component at $\omega = \pi B$, where B is the bit rate of the signal. Therefore, the transmitter power must be increased to compensate for the attenuation of the Fourier component which is measured in terms of dispersion penalty as

$$P_D \cong 10 \log_{10} \frac{1}{H(\pi B)} \cong 21(\sigma_p^2 B^2)$$ (2.41)

For example, to incur a dispersion penalty of 1 dB, the r.m.s pulse spread has to be about one fourth of the bit period ($\sigma_p = 0.25\ T$), where T is the period of the pulse, which is the inverse of the bit rate. Hence, the following condition can be used to estimate the dispersion.

$$B \le \frac{1}{4\sigma_P}$$ (2.42)

From (2.42), it can be inferred that the r.m.s impulse spread increases with the length of the fiber and hence it will limit the bit rate of the signal. Thus the condition given by (2.42) imposes the maximum length of the fiber and serves as a useful tool for comparing different lightwave systems. The maximum length that satisfies (2.42) is called the dispersion limited transmission distance.

2.1.3.2 Dispersion Mechanism in Fibers

The most important dispersive mechanism in multimode fibers is called the intermodal dispersion, which occurs due to the overlap of many propagating modes. An approximate estimate of this intermodal dispersion can be obtained in a step index fiber by referring to Figure 2.2. In this figure, it can be seen that the time taken by a typical ray corresponding to a specific mode to travel through a fiber of length L is

$$\tau = \frac{Ln_1}{c\,\cos\theta} \qquad 0 \le \theta \le \theta_C$$ (2.43)

where c is the free-space velocity of light and n_1 is the refractive index of the core.

If a narrow pulse is injected into a fiber then it will spreadout as it propagates along the fiber. The full r.m.s width of the broadened pulse is approximately given by [27]

$$2\sigma_P \cong \frac{Ln_1}{c}\left(\frac{1}{\cos\theta_C} - 1\right) \cong \frac{Ln_1\Delta}{c} \qquad (2.44)$$

Using (2.44) in (2.42) results in a dispersion limited BL product of 10 Mbps-km for a step index fiber having $n_1 = 1.46$ and $\Delta = 0.01$. However, the BL product can be increased by decreasing the intermodal dispersion by using the graded index fibers. In the case of graded index fibers, the transit time difference between the fastest and slowest propagating waves can be made approximately the same by carefully designing the index profile. Hence, the resulting r.m.s. pulse spread in graded index fibers is given by

$$\sigma_P = \frac{Ln_1\Delta^2}{8c} \qquad (2.45)$$

where n_1 is the index at the center of the fiber. Using (2.25) in (2.42), the dispersion limited BL product is given as

$$BL \leq \frac{2c}{n_1\Delta^2} \qquad (2.46)$$

Then BL products up to 2 Gbps-km can be easily achieved by using graded index fibers.

In the case of single mode fibers, since a single mode propagates, the intermodal dispersion is absent. However, the dispersive effects are not completely diminished. In this fiber, the time delay of the propagating mode is in general a function of signal wavelength. Therefore, each frequency component of the pulse travels with a different time delay. The time delay associated with these components is also called group delay. This group delay is given by [12]

$$\tau_D = \frac{L}{c}\frac{d\beta}{dk_0} = -\frac{\lambda^2 L}{2\pi c}\frac{d\beta}{d\lambda} \qquad (2.47)$$

where $k_0 = 2\pi/\lambda$ is used in (2.47). L is the fiber length, β is the propagation constant, and $c/d\beta/dk_0$ is called the group velocity of the propagating waves. The dispersion associated with this group velocity is known as group velocity dispersion or simply fiber dispersion. This dispersion has two components: (1) material or chromatic dispersion and (2) waveguide dispersion. Now, let us examine these two components. The material dispersion arises due to the wavelength dependence of the refractive index in a fiber, whereas the waveguide dispersion is characterized by the time required for a pulse to travel through a fiber length of L. Then, the total dispersion is made up of material and waveguide dispersion.

To estimate the total dispersion, let us consider a source having a spectral width $d\lambda$ and a fiber length L. Then the spectral components of a given mode will travel at different velocities and eventually the pulse output at the end of the fiber broadens. Hence, the increase in pulse width is characterized by [12]

$$\sigma_t = \frac{d\tau_D}{d\lambda} \sigma_\lambda \tag{2.48}$$

where σ_λ is the r.m.s. width which characterizes the source spectral width $d\lambda$, and $d\tau_D/d\lambda$ represents the delay difference, which can be further written as

$$\frac{d\tau_D}{d\lambda} = \frac{-L}{2\pi c}\left(2\lambda \frac{d\beta}{d\lambda} + \lambda^2 \frac{d^2\beta}{d\lambda^2}\right) \tag{2.49}$$

The term $(1/L)(d\tau_D/d\lambda) = D$ is the fiber dispersion measured in terms of picoseconds per kilometer-nanometer, which is the combination of the material and waveguide dispersion. These two components can be estimated separately as follows.

Let us assume a wave propagating in an infinitely extended dielectric medium that has a refractive index $n(\lambda)$. Using the propagation constant $\beta = 2\pi n(\lambda)/\lambda$ in (2.49) gives the group delay for material dispersion as

$$\tau_{DM} = \frac{L}{c}\left[n - \lambda \frac{dn}{d\lambda}\right] \tag{2.50}$$

Using (2.50), one can show that the pulse spread due to the wavelength dependency can be written as

$$\sigma_M = \frac{d\tau_{DM}}{d\lambda}\sigma_\lambda = -\frac{L}{c}\lambda \frac{d^2n}{d\lambda^2}\sigma_\lambda \tag{2.51}$$

The above equation represents the material dispersion, which is shown in Figure 2.10 as a function of wavelength. Note that the material dispersion is zero at 1.27 μm for a pure silica fiber.

Next, the waveguide dispersion can be estimated by assuming that the refractive index of the material is independent of wavelength. Let us consider a step index SMF with the normalized propagation constant b, which can be approximated as

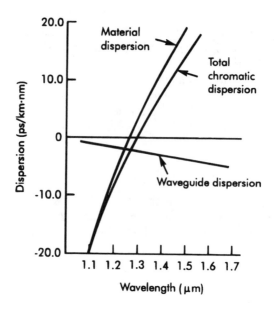

Figure 2.10 Material and waveguide dispersion in a single mode fused silica fiber. (*Source:* [28]. © 1979 IEE. Reprinted with permission.)

$$b = \frac{\left(\dfrac{\beta}{k_0} - n_2 \right)}{n_1 - n_2} \qquad (2.52)$$

solving (2.52) for β gives

$$\beta = k_0 n_2 (1 + b\Delta) \qquad (2.53)$$

Assuming that n_2 does not depend on the wavelength, we can estimate the group delay arising from the waveguide dispersion as

$$\tau_{DW} = \frac{L}{c} \frac{d\beta}{dk_0} = \frac{L}{c} \left[n_2 + n_2 \Delta \frac{d(bk_0)}{dk_0} \right] \qquad (2.54)$$

By using the approximation

$$V = k_0 a \sqrt{(n_1^2 - n_2^2)} \cong k_0 a n_2 \sqrt{2\Delta}$$

and writing the group delay in terms of V, we get

$$\tau_{DW} = \frac{L}{c} \left[n_2 + n_2 \Delta \frac{d(Vb)}{dV} \right] \cong n_2 \Delta \frac{d(Vb)}{dV} \qquad (2.55)$$

Now, the pulse spread due to waveguide dispersion is obtained by differentiating the group delay τ_{DW} with respect to the wavelength.

$$\sigma_W = \sigma_\lambda \frac{d\tau_{DW}}{d\lambda} = -\frac{n_2 L \Delta \sigma_\lambda}{c\lambda} V \frac{d^2(Vb)}{dV^2} \qquad (2.56)$$

The term

$$V \frac{d^2(Vb)}{dV^2}$$

can be estimated empirically to within five % accuracy by using [8]

$$V \frac{d^2(Vb)}{dV^2} = 0.8 + 0.549(2.834 - V)^2 \qquad 1.3 < V < 2.6 \qquad (2.57)$$

A plot showing both the material and waveguide dispersion components along with the total dispersion is shown in Figure 2.10.

The total pulse r.m.s spread in a step-index single mode fiber can be estimated by using

$$\sigma_P = L(\sigma_M + \sigma_W) \qquad (2.58)$$

Using (2.58) in (2.42), we get the dispersion limited BL product. The dispersion analysis of single mode fibers which employ arbitrary index profiles requires a tedious mathematical and numerical methods. These methods are described in [29–31]. The dispersion limited BL product in single mode fibers can be increased further by using the dispersion shifted or dispersion flattened fibers. The dispersion achieved in these fibers is shown in Figure 2.11. It can be seen that from Figure 2.11, the C-SMF and DS-SMF have a near zero dispersion at 1.3 and 1.55 μm respectively. On the other hand, the DF-SMF has a dispersion of less than 2 ps/nm-km over 1.3–1.6 μm wavelength region.

Another factor which limits the information capacity of the fiber is the polarization mode dispersion (PMD) which limits the BL product of a system. The PMD is a critical factor in the case of coherent detection schemes. This problem can be overcome by using the polarization maintaining fibers such as high birefringent fibers or single polarization single mode fibers [32–35]. The fiber medium is called birefringent, because there are actually two modes which are orthogonally polarized in a practical conventional single mode fibers and these modes propagate with different velocities. Therefore, the state of the polarization (SOP) of the light output varies randomly with time. Due to this birefringence in the fiber, the energy is coupled from

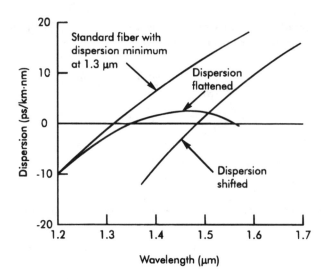

Figure 2.11 Dispersion curves for dispersion shifted and flattened single mode fibers. (*Source:* [122]. Reprinted with the permission of John Wiley and Sons.)

one mode to another. The polarization capacity of such a high birefringent fiber is measured by the beat length [7]

$$L_B = \frac{2\pi}{(\beta_x - \beta_y)} = \frac{\lambda}{B_f} \tag{2.59}$$

where β_x and β_y are the two propagation constants of the two orthogonally polarized modes in x and y directions respectively. B_f is called the birefringence of the fiber. From (2.59), it can be seen that a lower value of L_B gives higher value of birefringence, yielding higher polarization holding capacity at a given wavelength λ. In other words, with the introduction of high birefringence into a fiber, the SOP of the light output is maintained by decreasing the coupling between the two orthogonal modes. However, in this case, crosstalk could be a factor as the length of the fiber increases. Hence for longer distances, the fibers that support only one polarization mode are preferred.

In recent years, new techniques have emerged to combat the effect of chromatic dispersion in 1.55 µm conventional single mode fiber systems, which can be upgraded to avail the benefits of erbium doped fiber amplifiers for long-haul communications. A variety of these techniques is described in review articles in [36]. Some of the popular techniques include external modulation schemes, dispersion compensation fiber, dispersion supported transmission format, and soliton-based systems. The dispersion penalties that arise for the chirp of the direct modulation can be avoided by using an external modulator such as lithium niobate Mach-Zehnder modulator [37].

Another popular method is to add a dispersion compensation fiber that has a very large negative waveguide dispersion to a C-SMF system, so that the total fiber dispersion can be eliminated. A special dispersion compensation SMF has been realized at 1.55 μm [38,39].

In the case of a DST method [40], the dispersive fiber is used to convert the frequency modulated output signal into amplitude modulation, called the pure FSK transmission. The principle of DST method is described in Figure 2.12. The amplitude modulated signal is detected by a conventional optical receiver. In Figure 2.12(a), I denotes the transmitter driving signal, v is the optical frequency, and P_{opt} is the optical signal, which is constant. The optical frequency is switched by the incoming NRZ binary data with the frequency shift Δv, corresponding to the wavelength shift $\Delta\lambda = -\Delta v\lambda^2/c$, where c is the velocity of light. Because of the fiber dispersion, the various signal components with different wavelengths arrive at different times at the output of a fiber of length L. The resulting time difference is given by $\Delta\tau = \Delta\lambda DL$, where D is the fiber dispersion. The resulting amplitude modulation of the signal is shown in Figure 2.12(a). Note that P_{opt} is a three level signal but no longer a constant.

If the optical transmitter is modulated with a data signal of the bit rate B, then the three-level signal is detected by a receiver, whose integrated (V_{int}) and decision level (V_{dec}) outputs are depicted in Figure 2.12(b), which represents the original NRZ signal. The maximum achievable distance is given by [40]

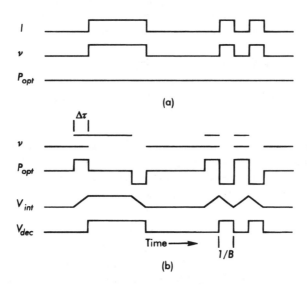

Figure 2.12 Principle of dispersion supported transmission: (a) transmitter signals and (b) receiver signals. (*Source:* [40]. © 1994 IEEE. Reprinted with permission.)

$$B^2L = \frac{2c}{D\lambda^2} \qquad (2.60)$$

where MSK (minimum shift keying) modulation is assumed with frequency shift $\Delta\nu = B/2$. Assuming $\lambda = 1.532$ μm, $D = 16.2$ ps/(nm.km), and $B = 10$ Gbps, the maximum achievable distance is 159 km, which is four times the conventional limit for external modulation. This method offers an advantage of direct modulation and simplicity, and it does not require additional components such as external modulators or dispersion compensating elements. Another technique uses the nonlinearity of the fiber to generate dispersion free pulses called solitons. This technique along with the concepts related to soliton transmission is discussed in Chapter 6.

2.1.4 Nonlinear Effects in Fibers

The information capacity of a lightwave system is ultimately limited by the nonlinear interactions between the information signals and the fiber medium. These optical nonlinear interactions can lead to interference, distortion, and attenuation of the signals, resulting in system degradation. There are four types of nonlinear effects in optical fibers: (1) stimulated Raman scattering (SRS); (2) stimulated brillouin scattering (SBS); (3) carrier-induced phase modulation; and (4) four photon mixing. These nonlinear effects degrade both single and multichannel systems in different ways. These effects are considered in a fused silica fiber. In the following paragraphs, we will discuss the nonlinear effects from the context of system limitations. Additionally, techniques to minimize the effects of nonlinearities are described.

2.1.4.1 Stimulated Raman Scattering

SRS is an interaction between the light and vibration of silica molecules. The incident light wave scattered by molecules undergoes a shift in the frequency. The change in frequency is called the Strokes frequency. Thus, if two optical waves separated by the Strokes frequency are co-injected into a Raman active medium, then the lower frequency (called the probe) wave gets amplified at the expense of the higher frequency wave (called the pump). The process of this amplification is called SRS. The amount of amplification in single mode fibers is given by [41]

$$P_S(L) = P_S(0)e^{(gL_eP/bA_e)} \qquad (2.61)$$

where $P_S(0)$ is the power of the incident wave, $P_S(L)$ denotes the power of the exiting or probe wave in a fiber medium of length L, P represents the injected pump power, and A_e denotes the effective area of the propagating waves. In general A_e is equal to the fiber core area provided that the pump and probe wavelengths are nearly the same, and g, the gain coefficient (cm/W), is a direct measure of the fiber nonlinearity. The factor b is related to the relative polarization of the waves and the polarization properties of the fiber. The term b is equal to 2 for the case of a conventional SMF and is equal to unity for the case of a polarization-maintaining fiber assuming identical polarization states for the pump and probe waves. The term $L_e = (1 - e^{\alpha L})/\alpha$ represents the effective fiber length, where α is the loss coefficient of the fiber. For $\alpha L << 1$, $L_e = 2$ and for $\alpha L << 1$, $L_e \cong 1/\alpha$. In a single channel lightwave system, the signal generates spontaneous Raman scattered light that can then be amplified. It has been shown that [41] the amplification of Raman-scattered light will cause system degradation when

$$\frac{gL_eP}{bA_e} = 16 \qquad (2.62)$$

For long fibers assuming $\alpha = 0.25$ dB/km, $A_e = 5 \cdot 10^7$ cm^2, $g = 7 \cdot 10^{12}$ cm/W, $b = 2$, the signal power required for system degradation is about 1.3 W and hence SRS will not be a problem in the case of single channel systems.

On the other hand, in the case of WDM systems, various channels at different wavelengths are injected into the fiber, and the signals at longer wavelengths will be amplified by the shorter wavelength signals, which leads to system degradation. In general, the interactions between the signals is more severe in the case of multiple channel systems. In this case, the following condition can be used to design a system for a 1 dB power penalty [41,42]

$$NP[(N - 1)\Delta f] < 500 \text{ GHz.W} \qquad (2.63)$$

where NP is the total optical power injected into the fiber, N is the number of channels in the system spaced at Δf apart, P is the power per channel, and $(N - 1)\Delta f$ represents the total occupied optical bandwidth. This result is of fundamental importance in designing multichannel WDM systems, which implies that the product of total optical power and total optical bandwidth must be less than 500 GHz-W to minimize degradation due to SRS. Hence, the effects of SRS cause a problem when hundreds of channels are multiplexed, and (2.63) serves as a guideline for the system design.

2.1.4.2 Stimulated Brillouin Scattering

SBS is somewhat similar to SRS, except that the SBS involves sound waves instead of molecules. As similar to SRS, the incident light is scattered into a light of longer wavelength. Hence, the scattered light is shifted to a lower frequency by an amount $f_B = 2nV_S/\lambda$, where n is the refractive index and V_S is the velocity of sound. At 1.55 μm, $f_B = 11$ GHz for silica fibers. SBS is much more nonlinear than SRS and it generates a potentially strong scattered beam and exhibits gain in the backward direction. SBS sets a limit on the injected power in lightwave systems. It is a dominant nonlinearity in single channel lightwave systems with a threshold power on the order of a few milliwatts. As transmitter powers increase to accommodate longer repeater spans, this effect will become an important factor in system design. The amount of nonlinearity depends on the Brillouin gain spectra of fibers, which depend on the fiber type and uniformity. The continuous wave (CW) threshold power for a uniform fiber at which SBS degrades a single channel system performance is given by [43]

$$P_{th} = \frac{42A_e}{g_0 L_e} \left(\frac{\Delta v_B + \Delta v_P}{\Delta v_B} \right) \tag{2.64}$$

where g_0 represents the SBS gain coefficient, which is determined by the fiber material parameters [43], Δv_B denotes the spontaneous Brillouin linewidth, which is typically 20 MHz for silica fibers at 1.55 μm and Δv_p represents the linewidth of the pump laser. For example, in a typical single mode non-dispersion shifted fiber, $A_e = 80 \ \mu m^2$, $\alpha = 0.22$ dB/km, $L_e = 19.74$ km, $g_0 = 4 \cdot 10^9$ cm/W, the SBS threshold is about 4.25 mW assuming $\Delta v_B / \Delta v_P << 1$.

In a multichannel system, it is shown that the threshold power remains constant with the increased number of channels [41]. Hence, the effects of SBS are directly related to the transmitted power levels but independent of the number of channels. However, the effects of SBS may degrade the passively multiplexed systems (nonfrequency selective elements such as directional coupler), employing fewer number of channels. Passive multiplexing of N channels by a star coupler, for example, reduces the power per channel by a factor N. The above conditions are valid only for the CW signaling conditions.

In practice, the threshold power depends on the modulation formats, since the SBS gain is a function of encoding schemes and on the ratio $B/\Delta v_B$, where B is the bit rate. In the case of ASK and FSK modulated systems, the SBS gain decreases by a factor of four with the increase in bit rate. Similarly, for high bit rate PSK systems, the SBS gain decreases linearly with the bit rate. The SBS threshold can be increased by employing higher linewidth lasers, which may be undesirable, because higher laser linewidth may degrade the dispersion characteristics.

2.1.4.3 Carrier-Induced Phase Modulation

In silica optical fibers, the refractive index changes with the light intensity level. The change in refractive index causes significant changes in the phase of the received optical fields. This type of nonlinearity, which affects only the phase of the propagating signal, gives rise to carrier-induced phase modulation (CIP). This effect can degrade the performance of lightwave systems that employ the phase modulation schemes.

In single channel systems, CIP is called self-phase modulation (SPM), and it converts the optical power fluctuations into phase fluctuations. The r.m.s phase variations due to the nonlinear refractive index is given by [41]

$$\sigma_\phi = 0.035\sigma_P \tag{2.65}$$

where σ_ϕ denotes r.m.s phase fluctuation in radians and σ_P is the r.m.s power fluctuation in mW. In practical single channel systems, the resultant phase noise is less than 0.04 radians, which corresponds to a power penalty of approximately 0.5 dB.

In multichannel systems, CIP is called the cross-phase modulation (CPM), due to the power fluctuations in other channels, in addition to self-phase modulation. The r.m.s phase fluctuations in a particular channel due to power fluctuations in other channels is given as [41]

$$\sigma_\phi = 0.07\sqrt{N}\sigma_P \tag{2.66}$$

where N is the number of channels. Much larger CIP is generated from residual AM when semiconductor lasers are directly phase modulated. Assuming a residual AM of 20% in semiconductor lasers, the maximum power per channel due to CIP for a 1 dB power penalty is given by [41].

$$P < \frac{21}{N} \tag{2.67}$$

where N denotes the number of channels. In practical multichannel systems, the effect of CIP is negligible. The CIP noise does not degrade the PSK systems employing external modulators.

2.1.4.4 Four-Photon Mixing

The change in the refractive index with the light intensity gives rise to four-photon mixing (FPM) or four-wave mixing [44–51] which is similar to intermodulation products in electrical systems. For example, two co-propagating waves at frequencies f_1 and f_2 mix and produce two frequencies at $2f_1 - f_2$ and $2f_2 - f_1$. Similarly, three co-propagating waves, generating nine new optical waves at frequencies $f_{ijk} = f_i + f_j - f_k$, where i, j, and k can be 1, 2, or 3. In general, for N channels launched, the number of mixing products generated is given by $M = 1/2(N^3 - N^2)$. Clearly, the generation of additional frequency components will degrade the multichannel system performance, through interchannel crosstalk. The effects of the four-photon mixing depends on the type of detection scheme (direct or coherent), channel spacing and the fiber dispersion. It has been shown that the FSK heterodyne envelope detection scheme is more sensitive to BER degradation then the FSK direct detection system [45,46]. This is attributed to the noise generated due to the beating between the local oscillator signal and the FPM signals. In addition, the relationship of the resulting power penalty due to FPM is different in these detection schemes [47]. The wider channel spacing decreases the efficiency of FPM. On the contrary, smaller fiber dispersion results in higher FPM efficiency. The methods to minimize the effects of these nonlinearities are discussed in Section 2.1.4.5.

2.1.4.5 *Minimization of Fiber Nonlinearities*

First, the effects of SRS in WDM systems can be minimized by restricting the product of total power and total optical bandwidth to be smaller than 500 GHz-W. Second, to combat the effects of FPM nonlinearity in WDM systems, several schemes have been proposed. In one scheme [49], the light emitted from a laser source is intensity modulated by an external modulator and at the same time the laser light frequency is sinusoidally modulated by injection current around the center frequency of each channel. Using this technique, the spectrum of the beat component between the selected signal and FPM lights is spread and the amount of beat component is reduced by using a baseband filter. In another scheme [50], some special devices such as optical multi-/demultiplexers were inserted into the transmission line to suppress the effects of FPM nonlinearity in multistage amplifier systems. The third scheme does not require any additional devices and it allows the use of a conventional transmitter. In this method [51], FPM nonlinearity effects have been minimized by spacing the channels unequally, ensuring that no mixing products fall on any of the channels, along with the use dispersion management. The dispersion management includes the use of a compensating fiber with large negative dispersion in conventional SMF systems or by using alternate segments of positive and negative dispersion fibers with equal length. Similarly the effects of CPM can be minimized by using a dispersion shifted fiber and conventional SMF [48].

SBS is not a problem in WDM systems, but it is sensitive to signal modulation format and is detrimental in systems employing narrow linewidth lasers. Hence, practical SBS suppression techniques should be designed for any fiber or any modulation technique. In one of the techniques described in [52], higher linewidths can be obtained by directly modulating a laser with a sinusoid at a frequency (called dither frequency) much lower than the low-frequency cutoff of the receiver. This will cause the laser to be FM modulated at a frequency that is outside the receiver bandwidth but will result in a large effective linewidth. To obtain complete SBS suppression through the entire effective length of the fiber, the dither frequency should not be above 6 kHz. The magnitude of the dither required for SBS suppression depends on the FM response of the laser. A 2.4% dither amplitude corresponds to a 528 MHz laser linewidth. Since the dither frequency is outside the receiver bandwidth, it will not degrade the signal in the presence of dispersion. In conclusion, SBS nonlinearity will not pose any system limitations if proper care is taken in the laser transmitter design.

2.2 LIGHT SOURCES

The generation of a modulated light output signal can be accomplished with a light source. Early stage development of semiconductor lasers and LEDs as light sources [53,54] paved the way for optical fiber communications. These light sources are compatible with the transmission characteristics and small physical dimensions of the low-loss optical fibers. The requirement of small physical dimensions was provided by sources that were fabricated from single crystals of group III-V compound semiconductors or their ternary or quaternary mixtures. The light sources can be categorized into short-wavelength and long-wavelength sources. The material GaAs-AlGaAs is used for the short-wavelength (0.7–0.9 μm) region and InP-InGaAsP material for the long-wavelength (1.3–1.6 μm) region.

The semiconductor lasers are based on coherent emission of light due to the stimulated recombination of the injected carriers by internal light. There are two types of semiconductor lasers: (1) multifrequency lasers and (2) single frequency lasers. The main difference between these lasers is their spectrum. The multifrequency lasers oscillate in several longitudinal modes simultaneously due to the small gain (or loss) difference between the adjacent modes, which results in many frequency components. Conversely the single frequency laser oscillates in a single longitudinal mode (dominant mode) and whereas the other modes are discriminated by their higher loses, resulting in a single frequency component. The semiconductor lasers offer higher power output, higher modulation bandwidth, and narrower spectral width than the LEDs. In contrast to lasers, LEDs make use of the incoherent light emission from the spontaneous emission of carriers. The LED device structure is compatible with multimode fibers and achieves higher light-coupling efficiency. In recent years, the light sources for optical fiber communications have been extensively reviewed in many excellent textbooks [55–59] and review articles [60–65].

In this section, we describe the basic operation, structures, and characteristics of semiconductor lasers and light emitting diodes for use in telecommunications. Section 2.2.1 discusses the basic concepts of these devices. Section 2.2.2 discusses the semiconductor lasers, and Section 2.2.3 describes the light emitting diodes. Section 2.2.4 compares LEDs with semiconductor lasers for system design applications.

2.2.1 Basic Concepts

The semiconductor lasers and LEDs use direct band gap semiconductors that belong to group III-V elements. The semiconductors for light sources are made from the compounds of group III and group V elements resulting in AlGaAs and InGaAsP materials. High enough radiative recombination of electrons and holes can only take place in direct band gap materials, which generates light. Many of the semiconductor lasers and LEDs use double heterojunctions to provide electrical and optical confinement. A good match between the crystal lattice parameters of the two adjoining heterojunctions is required to minimize any physical defects. The bandgap energy of these materials is related by

$$\lambda = \frac{1.24}{E_g} \qquad (2.68)$$

where λ is the emission wavelength in micrometers and E_g is the energy gap of the material in eV. The relationship between the energy gap and the crystal lattice constants for different III-V materials is shown in Figure 2.13. A heterojunction interface is created by choosing two material compositions having the same lattice constants, but with different energy band gaps. The energy gap of $Al_xGa_{1-x}As$ material resulting from the composition of GaAs that are lattice matched to AlAs can be found from [55]

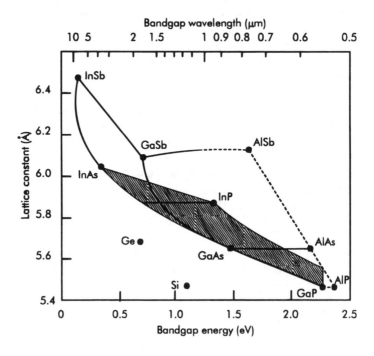

Figure 2.13 Energy gap versus the lattice constant of III-V materials. (*Source:* [57]. Reprinted with the permission of John Wiley and Sons.)

$$E_g = 1.424 + 1.266x + 0.266x^2 \qquad 0 \le x \le 0.37 \qquad (2.69)$$

where x is the ratio of AlAs to GaAs also called Al mole fraction. The value of x for the active material is normally chosen to yield an emission wavelength of 0.8–0.85 μm. For example with $x = 0.07$, the energy gap of $Al_{0.7}Ga_{0.93}As$ material is 1.51 eV, which emits light at $\lambda = 0.82$ μm. On the other hand, the band gap energy and lattice constant range for InGaAsP material is much larger than AlGaAs. The $In_{1-x}Ga_xAs_yP_{1-y}$ material is lattice matched to InP, whose band gap energy is given by [66]

$$E_g = 1.35 - 0.72y + 0.12y^2 \qquad (2.70)$$

where $y = 2x$ and $0 \le x \le 0.47$. For example, with $x = 0.26$, the alloy $In_{0.74}Ga_{0.26}As_{0.56}P_{0.44}$ has the energy gap of 0.96 eV, which can emit light at 1.3 μm.

The semiconductor lasers and LED's are fabricated by using epitaxial growth techniques on GaAs or InP. The three most commonly used techniques are: (1) liquid phase-epitaxy (LPE), (2) vapor-phase epitaxy (VPE) and (3) molecular-beam epitaxy (MBE). A detailed description of these fabrication methods is given in [56,57,59].

The basic operation of a light source involves three processes: (1) absorption; (2) spontaneous emission; and (3) stimulated emission. These three processes are described in Figure 2.14. When a photon energy $h\nu$ of the incident light falls on the system, an electron in the valence band will absorb the photon and then eventually reach the conduction band, which is shown in Figure 2.14(a). This absorption process occurs if the photon energy $h\nu = E_c - E_v$, where E_c is the energy of the conduction band, E_v is the energy of the valence band, and $\nu = c/\lambda$ is the frequency. In this case there is no light emission. The spontaneous emission of the light occurs only if the electron returns to the valence band, which is the basic principle of operation of an LED. In this case, the photons are emitted in random directions without any phase difference, and the emitted light is incoherent. The spontaneous emission process is illustrated in Figure 2.14(b).

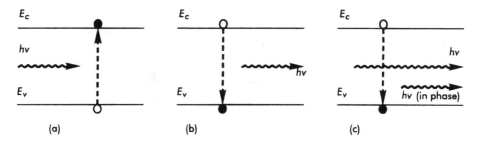

Figure 2.14 Basic operation of light generation: (a) absorption, (b) spontaneous emission, and (c) stimulated emission.

The laser action requires the stimulated emission, shown in Figure 2.14(c). In this case, when the incident light of photon energy $h\nu$ falls on the system, an electron from the conduction band is stimulated to drop to the valence band by means of population inversion, results in coherent light. The emitted photon is in phase with the incident photon. The population inversion in semiconductor lasers occurs if the population of the conduction band is greater than that of the valence band, a result that can be achieved by injecting electrons into a heavily doped active layer of the laser.

2.2.2 Semiconductor Lasers

In principle, the semiconductor laser is an optical waveguide with reflecting facets at each end that form a resonant cavity. The basic structure of a semiconductor laser is shown in Figure 2.15. The light is confined and guided by a three-layer dielectric waveguide structure called a Fabry-Perot cavity [55]. The refractive index of the active layer is chosen to be higher than its surrounding layers so that the propagation of the electromagnetic radiation is guided in a direction parallel to the layer interfaces.

The basic operation of a laser can be understood by referring to Figure 2.16. The planes at $y = 0$ and $y = d$ are GaAs-AlGaAs heterojunctions, one of which is the injecting p-n junction. The p-n junction (part of double-heterojunction), which is formed by adjoining p and n-type layers, is shown in Figure 2.16(a). A narrow depletion region (no mobile carriers) exists at the junction. Its width depends on the applied voltage and carrier concentration in the p- and n-regions. As the applied voltage is raised, the potential barrier across the junction is reduced and a net flow

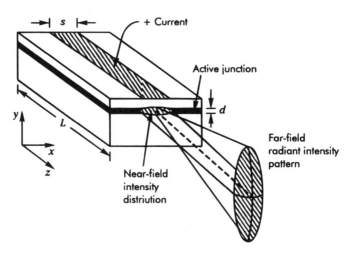

Figure 2.15 Illustration of a basic semiconductor laser, showing the far and near field radiation patterns. Here, s denotes the stripe width, d is the thickness of the active region, and L is the length of the cavity. (*Source:* [8]. Reprinted with the permission of Irwin, Inc.)

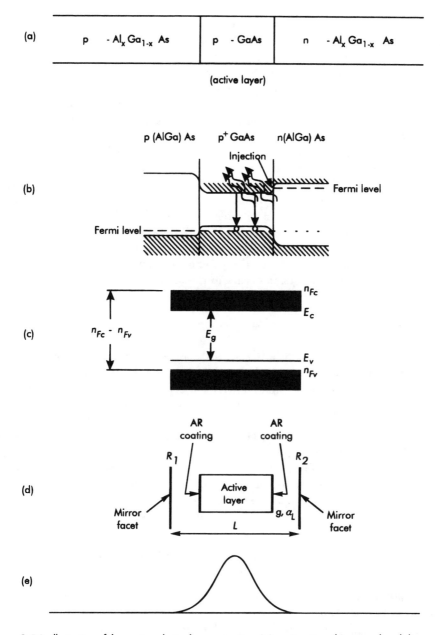

Figure 2.16 Illustration of the semiconductor laser operation: (a) *p-n* junction; (b) energy band diagram; (c) population inversion region; (d) lasing action; and (e) light intensity profile.

of carriers across the junction results. The magnitude of the potential barrier depends on the initial positions of the Fermi levels relative to the band gap energy. The energy band diagram of the p-n junction under forward biased condition is shown in Figure 2.16(b). If the applied voltage is large enough, then the electrons are injected into the p-region and holes into the n-region across the transition region. This region contains a large concentration of electrons within the conduction band and holes within the valence band. If these population densities are large enough, a condition of population inversion results in this region.

The quasi-Fermi levels shown in Figure 2.16(b) show the deviation from equilibrium (no applied voltage) caused by the applied voltage. When a large number of electrons are injected across the junction, the electron concentration begins at a maximum value near the junction and decays exponentially to its equilibrium value in the p-material. A similar argument holds for the holes. The separation of quasi-Fermi levels at any point is a measure of the deviation from the equilibrium, and this deviation is quite significant in the vicinity of the junction. The expanded view of this deviation at the vicinity of the junction is shown in Figure 2.16(c). The condition of population inversion occurs in the laser when the difference of quasi-Fermi levels is greater than the photon energy emitted, that is

$$\eta_{FC} - \eta_{FV} > h\nu \tag{2.71}$$

where η_{FC} and η_{FV} are the quasi-Fermi levels (represent the carrier concentration) in the conduction and valence bands respectively. For band-to-band transitions, the minimum requirement for population inversion occurs when $h\nu = E_g$ in (2.71).

The lasing action in a Fabry-Perot cavity can be understood by referring to Figure 2.16(d), where the laser cavity is of length L with two mirror facets of reflectivity R_1 and R_2, respectively. The mirror facets are formed by cleaving the two ends of the laser. The terms g and α_L represent the gain and internal losses in the laser cavity respectively. The lasing action occurs when the optical gain matches with the losses encountered over a round trip of the beam through the cavity, which is the threshold condition for lasers. Therefore, the threshold condition for lasing action is given by [55]

$$g = \alpha_L + \frac{1}{2L} \ln\left(\frac{1}{R_1 R_2}\right) \tag{2.72}$$

The second term in (2.72) denotes the useful laser output. The threshold current density of the laser is given by $J_{TH} = g/K_1$, where $K_1 = 0.015$ is a proportionality constant for double heterostructure lasers. The resulting intensity distribution of the coherent light is shown in Figure 2.16(e).

The coherent light of radiation within the resonance cavity of a semiconductor laser gives rise to a pattern of electric and magnetic fields called modes. These modes can be classified into transverse electric (TE) and transverse magnetic (TM) modes. These modes can be described in terms of lateral, longitudinal, and transverse directions of the laser cavity. The longitudinal modes determine the frequency spectrum of the emitted light, which, in turn, depends on the length of the cavity. The lateral modes determine the shape of the lateral profile of the laser beam, which is governed by the lateral structure of the device. The transverse modes determine the radiation pattern and the threshold current density, which are governed by the dimensions of the active region. Hence, the optical and electrical confinement mainly depends on the type of laser geometry and its structure.

2.2.2.1 Laser Structures

In the previous paragraphs, the semiconductor laser was modeled as a simple Fabry-Perot structure. In a broad area laser such as a Fabry-Perot structure, the lateral current is poorly controlled, and several lateral modes can oscillate simultaneously in addition to longitudinal modes. The longitudinal modes result from the Fabry-Perot resonances of the cavity formed by the cleaved laser mirrors. Consequently, the emitted light spreads over the entire width of the laser. Hence, the frequency spectrum of the beam becomes complex and unstable, leading to pulse spreading. In addition, the threshold current to drive the laser becomes very large due to the larger area. Therefore, the lasers have to be designed for good lateral current and mode confinement.

   In the following paragraphs, we will outline some of the practical laser structures and describe some of their important characteristics. The practical laser structures can be broadly classified into two types: (1) gain-guided or stripe geometry lasers and (2) index-guided lasers. In the case of gain-guided lasers, the optical guidance is mainly gain dependent. In these lasers, the optical gain peaks at the center of the stripe; hence, the light is confined to the stripe region. The lateral field distribution is controlled and the threshold current of the laser is decreased by an order of magnitude. Some of the typical gain-guided laser structures [67] are shown in Figure 2.17. The lateral control of the current and light can be achieved by utilizing either the oxide type or zinc diffusion type of stripe laser. The cross-sectional views of these laser structures are shown Figures 2.17(a) and 2.17(b), respectively. Because of the narrow metal contact, these structures provide better current injection and hence result in lower threshold currents. With $s = 10$ μm, these lasers can be coupled to multimode fibers with higher coupling efficiency. Conversely, these lasers suffer from kinks (nonlinear light vs. current characteristics) and self-pulsations. Also, unstable far-field patterns may result due to the jumps between the sets of lateral modes.

   The problems of kinks, self-pulsations, and unstable field patterns can be reduced by the use of index-guided laser structures. In the case of index-guided structures, a discontinuous change in the refractive index (step index Δn) is built into the structure in the lateral direction hence the optical confinement is index

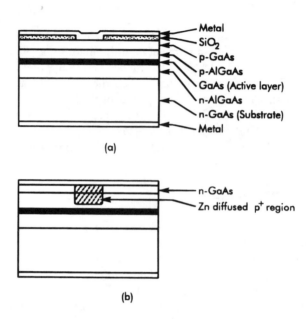

Figure 2.17 Gain-guided lasers: (a) oxide type and (b) zinc diffusion type. (*Source:* [8]. Reprinted with the permission of Irwin, Inc.)

dependent. Depending on the magnitude of Δn, the index-guided structures can be classified into two types: (1) weak index guiding structures; and (2) strong index guiding structures. In the case of weak index guiding, the active region waveguide thickness is varied by growing it over a channel in the substrate as depicted in Figure 2.18(a). This structure is called channel-substrate planar waveguide laser [68]. This type of structure requires higher threshold currents and yields output powers up to 20 mW. Another weak indexing structure called the ridge-waveguide laser is shown in Figure 2.18(b). In this structure, the waveguide provides a narrow current confinement in addition to the weak index guiding. This type of structure yields a threshold current of 18 mA with a power output of 25 mW at 1.5 mm [69].

In the case of strong index guiding lasers, excellent current as well as optical confinement results. Examples of this type are etched mesa buried heterostructure and channel substrate buried-heterostructure lasers [70–72]. In these structures, the active volume is buried in a material of wider band-gap and lower refractive index. Room temperature threshold currents of 15–25 mA are typical with the power outputs of 30–40 mW. These structures are shown in Figure 2.18(c,d).

The index-guided laser structures provide a good control for lateral modes. However, several longitudinal modes can also oscillate in these devices, a situation that results in many frequencies and leads to pulse spreading during propagation through the fiber, and in addition results in mode partition noise (due to fluctuations in the modal distribution from pulse to pulse). The frequency spacing (Δf) between

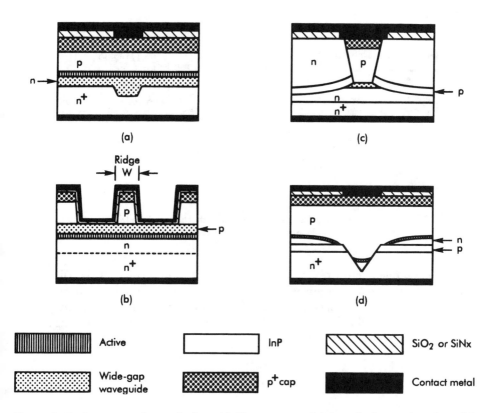

Figure 2.18 Cross-sectional view of index-guided laser structures: (a) channel-substrate planar laser; (b) ridge-waveguide laser; (c) etched mesa buried-heterostructure laser; and (d) channel substrate buried-heterostructure laser. (*Source:* [13]. Reprinted with the permission of Academic Press.)

successive modes is given by $c/2Ln$, where c is the free space velocity, L is the length of the laser cavity, and n denotes the refractive index. This can be avoided by operating the laser in a single frequency or single longitudinal mode (SLM). These lasers are known as single frequency or dynamic single mode lasers.

The principle involved with the single frequency lasers is to provide adequate gain or loss discrimination between the desired mode and all of the unwanted modes of the laser cavity. After years of development in this area, single frequency operation was achieved by using Bragg reflectors or gratings, which provide optical feedback for lasing action instead of cleaved facets. These structures are known as distributed feedback (DFB) or distributed Bragg reflector (DBR) lasers. These structures are shown in Figure 2.19. In the case of a DFB laser structure, the grating or the corrugated region is built into the pumped active region of the laser. The feedback occurs based on the Bragg diffraction principle, which couples the forward and backward propagating waves. In the case of a DBR laser, the feedback does not take place

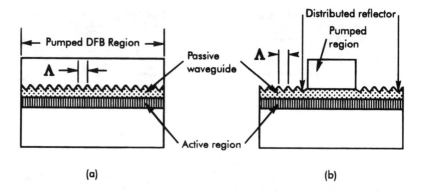

Figure 2.19 Cross-sectional view of single frequency laser structures: (a) the DFB laser and (b) the DBR laser. (*Source:* [13]. Reprinted with the permission of Academic Press.)

inside the pumped active region but results from the end regions of the DBR laser. The DFB lasers have been utilized for direct modulation of the laser current up to 10 Gbps in recent experiments at 1.5 μm [72]. A detailed analysis of the DFB lasers has been carried out in [73,74]. The SLM operation is achieved if the grating spatial period L is chosen as

$$\Lambda = \frac{L\lambda_B}{2n} \tag{2.73}$$

where λ_B is the operating Bragg wavelength, L is the length of the cavity, and n is the refractive index of the grating medium. The best structure for a stable single frequency operation is the quarter-wave shifted DFB laser [75]. In this structure, a $\pi/2$ phase shift is incorporated in the grating at the center of the laser cavity with both ends antireflection coated. When the phase shift is $\pi/2$, the lowest threshold gain is obtained for the main mode exactly at the Bragg wavelength and hence results in a large gain difference between the main mode and the side modes. The performance of this structure is superior to that of conventional DFB lasers in terms of dynamic single-mode stability and negligible mode partition noise at multigigabit speeds. Additionally, a narrow linewidth has been obtained under cw operation. Prior to the development of DFB lasers, other laser structures such as coupled cavity and external cavity laser structures were demonstrated for SLM operation [76,77], but these are difficult to realize for commercial applications.

The next generation of semiconductor laser structures falls into quantum wells [78–81]. These structures offer more than an order of magnitude reduction in threshold current and a much narrower spectral width than the conventional semiconductor lasers. In a quantum well laser structure, the thickness of the active region is much smaller than that of a conventional laser, so that the electronic and optical properties change drastically. The quantum well lasers can be broadly classified into several types: (1) single quantum well lasers; (2) multiple quantum well lasers or superlattice structures; (3) quantum wire structures; and (4) quantum dot structures. The single quantum well laser has one thin active layer less than 10 nm as compared to 100 nm for a double heterostructure (DH), whereas the multiple quantum well lasers have several thin active layers. In the case of a quantum wire structure, the laser takes the form of a thin wire of rectangular cross section, and the carriers are confined in all three directions within a box in a quantum dot structure. The implication of laser characteristics is quite significant. First, the reduced density of states requires fewer electrons to reach the lasing threshold, resulting in lower threshold current. Second, the differential gain is much larger than that of bulk materials, giving rise to narrower linewidths. The threshold currents of less than 1 mA and linewidths of less than 10 MHz can be achieved in some practical devices.

2.2.2.2 Semiconductor Laser Characteristics

Analysis of semiconductor lasers plays an important role in understanding their characteristics, and it is useful for optimizing their performance and efficiency. For this purpose, several models have been developed for describing semiconductor lasers' electrical, optical, and thermal characteristics [82–90]. In this section, we review some of the important characteristics of semiconductor lasers. The power output of the laser inside a laser cavity due to an applied current I can be expressed as [59]

$$P_o = \frac{h\nu}{q} \left[\eta_i (I - I_{Th}) \right] \qquad I > I_{Th} \qquad (2.74a)$$

$$P_o = \frac{h\nu}{q} \eta_i I \qquad I < I_{Th} \qquad (2.74b)$$

In (2.74a), the term $(I - I_{Th})$ represents the number of injected electrons, η_i is the internal quantum efficiency of the laser, which is a measure of the total photons emitted inside the laser cavity per second, and I_{Th} is the threshold current density, which is temperature dependent, given by [91]

$$I_{Th} = I_{Th}(0) \exp(\Delta T / T_C) \qquad (2.75)$$

where $I_{Th}(0)$ is the current at a reference temperature, ΔT is the temperature difference between the laser and the reference value, and T_C is the characteristic temperature of the laser. T_C ranges from 100 to 200K for AlGaAs lasers and 40 to 80K for InGaAsP lasers. Hence, InGaAsP lasers have much stronger dependence on temperature than AlGaAs lasers. Some part of the power given in (2.74a) is coupled out through the mirror facets as a useful power output. Using (2.72), the actual power output of the laser is given by

$$P_O = \frac{hv}{q} \left[\eta_i (I - I_{\mathrm{Th}}) \right] \frac{\dfrac{1}{2L} \ln\!\left(\dfrac{1}{R_1 R_2} \right)}{\alpha_L + \dfrac{1}{2L} \ln\!\left(\dfrac{1}{R_1 R_2} \right)} \tag{2.76}$$

The third term in (2.76) represents the ratio of the power radiated through the mirrors to the total power generated by the laser cavity. The typical power output vs. current (*P-I*) characteristics for a semiconductor laser is shown in Figure 2.20 at different temperatures. From Figure 2.20, it can be seen that when $I < 12$ mA (for the curve at 30°C), only the spontaneous emission is radiated, which resembles the *P-I* characteristics of an LED as per (2.74b). When $I > 12$ mA, the laser output increases significantly due to the stimulated emission. Also, it can be inferred that the laser threshold current increases with an increase in temperature resulting in a lower optical power output.

    The external differential quantum efficiency of the laser is defined as the number of photons generated for each electron-hole pair injected into the *p-n* junction, above

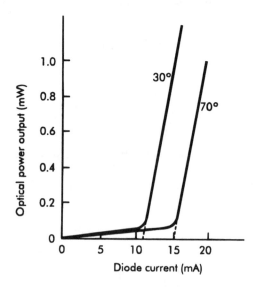

Figure 2.20 *P-I characteristics of a typical semiconductor laser as a function of temperature.*

the threshold current. Experimentally, it can be calculated from the straight line approximation of the *P-I* curve as

$$\eta_{ext} = \frac{q\Delta P}{hv\Delta I} \tag{2.77}$$

where ΔP is the incremental change in the emitted power over the corresponding change in the current ΔI. The external quantum efficiency varies by 30–40% for a typical semiconductor laser.

One of the most important characteristics of a laser diode is its modulation characteristics. The modulation characteristics of a laser diode depend on the device geometry and structure. These modulation characteristics should be determined under the pulse and CW modulation. We use the method described in [92,93] to derive the modulation bandwidth under pulse and small-signal CW modulation. The rate of change of carrier density in the laser cavity is given by

$$\frac{dn(t)}{dt} = \frac{J}{qd} - \frac{n(t)}{\tau} - C[n(t) - n_m]S(t) \tag{2.78}$$

where $n(t)$ is the carrier density, J is injected the current density, d is the active layer thickness, q is the electronic charge, $S(t)$ is the photon density, and τ represents the carrier lifetime, which is given by $\tau = [\tau_r\tau_{nr}]/[\tau_r + \tau_{nr}]$ where τ_r and τ_{nr} are the radiative and nonradiative carrier lifetimes respectively. The term $C(n(t) - n_m)$ in (2.78) denotes the optical gain in the laser cavity, where C is a constant and n_m represents the minimum carrier density required to achieve a positive gain. Similarly, the rate of change of photon density in the laser cavity is given by

$$\frac{dS(t)}{dt} = C[n(t) - n_m]S + \Gamma\frac{n(t)}{\tau} - \frac{S(t)}{\tau_P} \tag{2.79}$$

where Γ denotes the fraction of the spontaneous emission and τ_P is the photon lifetime. Equations (2.78) and (2.79) are called the coupled rate equations. First, let us evaluate the response of a laser for a current pulse modulation, which is shown in Figure 2.21(a). Prior to the laser diode being switched ON by the current pulse $J = J_0u(t)$, the number of photons in the cavity region is small, and hence the third term in (2.78) can be neglected. Then, (2.78) can be written as

$$\frac{dn(t)}{dt} = \frac{J_0u(t)}{qd} - \frac{n(t)}{\tau} \tag{2.80}$$

where $u(t)$ is the unit step function. Equation (2.80) can be solved, subject to the boundary condition $n(O) = 0$. Hence, the solution of $n(t)$ is

$$n(t) = \frac{\tau J_0}{qd}(1 - e^{-t/\tau})u(t) \qquad (2.81)$$

For $J > J_{Th}$, let $n(t) = n_0$ at $t = t_d$, then the carrier density is given by

$$n_0 = \frac{\tau J_0}{qd}(1 - e^{-t_d/\tau}) \qquad (2.82)$$

from which an expression for the delay time is approximately given by

$$t_d = \tau \ln \frac{J_0}{J_0 - J_{Th}} \qquad (2.83)$$

where J_0 is the bias current and J_{th} denotes the threshold current. The transient analysis of the laser results in a damped oscillatory output as shown in Figure 2.21(b). In essence, the system behaves like a tuned circuit at some resonant frequency. These are known as relaxation oscillations and set a upper limit to the direct current modulation frequency. This resonant frequency is given by [59]

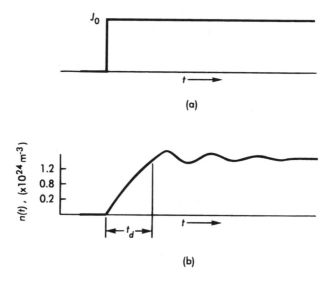

Figure 2.21 Laser response under pulse modulation: (a) current pulse and (b) laser response for $\Gamma = 10^{-4}$. (*Source:* [92]. ©1975 IEE. Reprinted with permission.)

$$f_r \cong \frac{1}{2\pi} \sqrt{\frac{1}{\tau \tau_P}\left(\frac{J}{J_{\text{Th}}} - 1\right)} \tag{2.84}$$

where τ_P is the photon lifetime. It can be seen that the resonance frequency increases with the decrease of carrier and photon lifetimes as well as with the increase of J/J_{Th}. Also, the modulation index is maximum at the resonant frequency and it decreases above the resonant frequency. The resonant frequency of some of the high performance lasers can be as high as 30 GHz.

Next, let us look at the small-signal cw modulation of the laser diodes. Let us assume a high-frequency sinusoidal current drive of the form $J = J_0 e^{j\omega t}$, with a signal frequency ω. To examine the frequency response of the laser, let us write the expressions for $J(t)$, $n(t)$, and $S(t)$ in terms of dc and ac components.

$$J(t) = J_0 + J_1 e^{j\omega t} \tag{2.85}$$

$$n(t) = n_0 + n_1 e^{j\omega t} \tag{2.86}$$

$$S(t) = S_0 + S_1 e^{j\omega t} \tag{2.87}$$

where J_0, n_0, and S_0 are the dc components and $J_1 e^{j\omega t}$, $n_1 e^{j\omega t}$, and $S_1 e^{j\omega t}$ represent the corresponding ac components. Substituting (2.85)–(2.87) into (2.78) and (2.79), it can be shown that (see Problem 2.20) the frequency response is given by the ratio of photon density to the injected current density as

$$\left|\frac{S_1}{J_1}\right| = \frac{\tau_P \omega_0^2 / qd}{\left[(\omega_0^2 - \omega^2) + \omega^2\left(\dfrac{1}{\tau} + \tau_P \omega_0^2\right)\right]^{1/2}} \tag{2.88}$$

where

$$\omega_0 = \sqrt{\frac{1}{\tau \tau_P}(1 + C n_m \tau_P)\left(\frac{J_0}{J_{\text{Th}}} - 1\right)} \tag{2.89}$$

which gives the modulation bandwidth of the laser. As seen from (2.89), the modulation response of the laser depends on $\tau \tau_P$ and J_0/J_{Th}. The frequency response of a directly modulated laser is plotted in Figure 2.22 for different values of τ/τ_P and J_0/J_{Th}. As can be seen from Figure 2.22, the modulation bandwidth of the laser increases as the photon lifetime is decreased or when the drive current is increased. The typical modulation bandwidth is on the order of tens of GHz.

Another parameter associated with the direct modulation of laser diodes is the frequency chirp. The amplitude modulation of the laser also results in a phase modulation, causing the laser mode frequency shift, which is known as frequency chirp. The frequency chirp can be approximated by [94]

$$\Delta v(t) = -\frac{\alpha}{4\pi} \left[\frac{d}{dt} \ln P(t) + kP(t) \right] \tag{2.90}$$

where α denotes the ratio of the change of index to the change of gain with respect to carrier change, P is the optical power, and $k = 2\Gamma\epsilon/V\eta h\nu$. Γ is the optical confinement factor, V is the volume of the active layer, ϵ is the gain saturation parameter, and η is the external quantum efficiency. From (2.90), it can be inferred that the

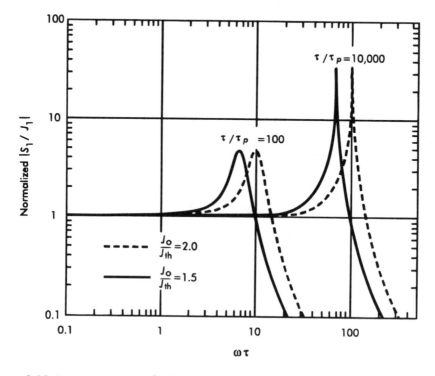

Figure 2.22 Frequency response of a directly modulated laser. (*Source:* [93]. ©1970 IEEE. Reprinted with permission.)

frequency chirp arises from two factors. The first term denotes the transient chirp, arises from the relaxation oscillation during the onset of the laser pulse. The second term represents the adiabatic chirp that stems from the damping of the relaxation oscillating frequency, leading to wavelength shifts of the optical waveform. The wavelength shift broadens the pulse spectrum of the laser output, which can limit the performance of directly modulated systems.

For a communication system designer, the noise characteristics of semiconductor lasers is very important. The noise characteristics of semiconductor lasers are: (1) spectral linewidth; (2) intermodulation distortion; (3) relative intensity noise; (4) modal noise; and (5) partition noise. The spectral linewidth characteristics are important for coherent lightwave system applications. The spectral linewidth in a laser is directly related to the phase fluctuations of the optical field. The phase fluctuations arise because of the spontaneous emission noise. The spectral linewidth Δv and the power output P_0 are related by [95]

$$\Delta v = \frac{v_g^2 h v g n_{sp} \alpha_m (1 + \alpha_l^2)}{8\pi P_0} \tag{2.91}$$

where g is the laser gain; n_{SP} is the spontaneous emission factor, which varies from 2 to 3; v_g is the group velocity; α_m represents the mirror loss; and α_l is the linewidth enhancement factor, which ranges from 2 to 16, depending on the material composition and wavelength. It can be seen that as the laser power increases, the laser linewidth decreases, because spontaneous emission becomes less significant at higher photon densities. Also, the linewidth decreases with the increase in laser length because the effective mirror loss per unit length decreases. Linewidths in the kilohertz range have been obtained by using external cavity and quantum well lasers and linewidths of few megahertz can be achieved with DFB lasers. The resulting spectrum of the phase noise is approximated by a Lorentzian shape, which is given by [96]

$$S(f) = \frac{P_0}{1 + \left[\dfrac{2(f - f_0)}{\Delta v}\right]^2} \tag{2.92}$$

The intensity fluctuations of the laser output are measured by the relative intensity noise (RIN). It is a ratio of the total mean-square noise power (p_n^2) per unit bandwidth (B) to the mean optical output of the laser (p^2) as [97]

$$\text{RIN} = \frac{\langle p_n^2 \rangle}{p^2 B} \tag{2.93}$$

RIN is measured in terms of decibels per hertz. RIN decreases with increase in the mean optical power output. Typical RIN for a semiconductor laser is -156 dB/Hz at 1 mW power. Another important factor that arises in analog modulation systems is the intermodulation distortion due to the nonlinearity of the laser L-I characteristics. Measurements and analysis [98] have shown that second harmonic distortions of lower than −60 dB can be obtained in directly modulated lasers at the low-frequency range. However, if the modulation frequency increases above 1 GHz, the second and third harmonics can be as high as −15 dBc (15 dB down relative the carrier) at a moderate optical modulation depth.

Other significant problems associated with the laser linearity are modal or speckle noise and reflection noise. The modal noise is due to the intensity fluctuations of the longitudinal modes of the laser, which particularly occurs in multimode lasers. As a result of light output fluctuations, the signal arrival times at the receiver can vary as the laser mode output pattern changes, which can degrade the performance of the receiver. The reflection noise is due to the optical feedback from the fiber joints into the laser cavity, which can affect the laser spectrum. The reflection problems can be avoided by using optical isolators between the laser diode and the optical fiber. The mode partition noise arises due to the propagation differences among many longitudinal modes propagating in a single mode fiber. This problem can be avoided by using the wavelength control techniques at the receiver.

2.2.3 Light Emitting Diodes

The LED is an incoherent light source that emits light by spontaneous emission. The spontaneous emission process requires lower threshold currents than the lasers. Also, an LED does not require an optical cavity and mirror facets to provide optical feedback. The LED propagates many modes and is thus called a multimode source, making it suitable for use with multimode fibers. The LEDs are more reliable and cost-effective than the semiconductor lasers. The disadvantages of LEDs are the low power output and smaller modulation bandwidth. However, in recent years, superluminescent LEDs have power outputs comparable to that of semiconductor lasers.

The first generation of LED systems used AlGaAs sources in the 0.8–0.9 μm region. These systems are primarily used for short haul links with a moderate bit rate up to 50 Mbps, due to the high-loss and dispersion of fiber. The second generation LED systems utilized the long-wavelength InGaAsP sources at 1.3 μm wavelength, where silica fibers have low loss and minimum dispersion. These systems yield a repeater spacing of tens of kilometers with a bit rate in the hundred megabits per second range [99]. The primary applications of LEDs are as short-haul data links, intra-city trunks, and loop feeders. Recent experiments have shown that 1.3 μm LED systems can launch high power into single-mode fibers for subscriber loop applications at bit rates as high as 560 Mbps [100].

2.2.3.1 *LED Structures*

The double heterojunction LED structures can be broadly classified into two types: (1) surface-emitting structures and (2) edge-emitting structure. The schematic diagram of a typical surface-emitting LED is shown in Figure 2.23.

This structure has been designed to obtain high radiance and couple the light efficiently to a fiber. The photons are generated from the thin *p*-GaAs active layer, *p*-AlGaAs and *n*-AlGaAs are the cladding regions for the light confinement. The top as well as the bottom *p*-GaAs layers are used for the ohmic metal contacts. The thin SiO$_2$ layer isolates the contact layer from the heat sink. The fiber is aligned with the light emitting region of the LED for achieving high coupling efficiency. However, the coupling efficiency is usually low due to the Lambertian distribution of the radiation intensity. The coupling efficiency can be improved by using an edge emitting LED. The geometry of a typical edge emitting LED is shown in Figure 2.24. This structure

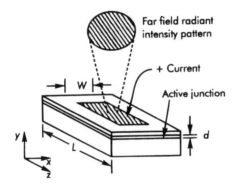

Figure 2.23 Surface-emitting LED structure. Here, *d* is the thickness of the active region, *L* is the laser length, and *W* is the width of the emitting region. (*Source:* [8]. Reprinted with the permission of Irwin, Inc.)

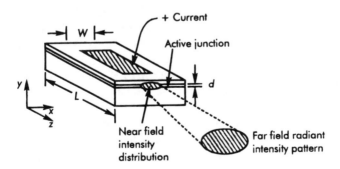

Figure 2.24 Edge emitting LED structure. (*Source:* [8]. Reprinted with the permission of Irwin, Inc.)

is similar to that of a stripe-geometry semiconductor laser. The waveguiding and reduced beam divergence gives improved light coupling efficiency, which is particularly important for single mode fibers having a small acceptance angle.

2.2.3.2 LED Characteristics

The light-current characteristics of an edge emitting LED along with its intensity patterns is shown in Figure 2.25. The optical power generated in an LED by the radiative recombination of electrons and holes is given by (2.74b), which in principle implies that the output power is directly proportional to the drive current for the given frequency and quantum efficiency. However, in practice, an LED can exhibit some degree of nonlinearity as shown in Figure 2.25(a), depending on the material properties and device geometry. The spectral linewidth of an LED is measured in terms of full width at half maximum (FWHM) and is typically about 40 nm, which is broader than that of semiconductor lasers.

The modulation bandwidth of an LED is mainly determined by its frequency response. When an LED current is modulated at an angular frequency w, then the intensity of the light output will vary with ω as [101]

$$|I(\omega)| = \frac{I(0)}{\sqrt{1 + \omega^2 \tau^2}} \tag{2.94}$$

where $I(0)$ is the intensity at zero modulation frequency and τ is the total recombination lifetime, which is approximately equal to the radiative recombination lifetime τ_r. The associated modulation bandwidth is measured by the cut-off frequency at which the detected power is one-half that at zero modulation frequency as

$$f_{3dB} = \frac{1}{2\pi\tau_r} \tag{2.95}$$

It is clear that the modulation bandwidth of the LED depends on the radiative carrier lifetime, which in turn depends on the device geometry. The modulation bandwidths of LEDs are much smaller than those of semiconductor lasers. Nevertheless, modulation bandwidths over 1 GHz can be achieved in some practical LEDs.

A third device known as superluminescent LED offers a higher power output, a narrower beam, and a smaller spectral bandwidth than the edge emitting LED [102]. The structure is similar to a laser device, in which the spontaneous emission experiences stimulated gain over an extended path with one mirror end, without any feedback. However, these devices require much higher threshold currents than the lasers. Nevertheless, recent improvements in lasers have made the superluminescent diodes more practical. Table 2.1 gives a comparative summary of the basic parameters for some of the practical edge emitting LEDs and DFB semiconductor lasers.

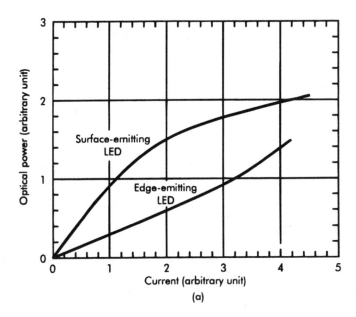

(a)

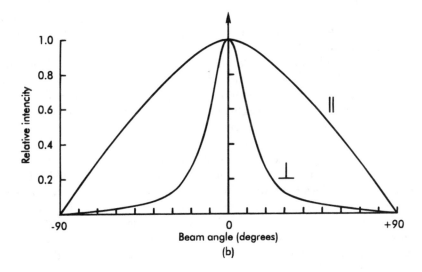

(b)

Figure 2.25 (a) Light current characteristics of surface-emitting and edge-emitting LEDs and (b) Intensity patterns of an edge-emitting LED in the planes normal and perpendicular to the junction. (*Source:* [8]. Reprinted with the permission of Irwin, Inc.)

Table 2.1
Comparison of Some Practical Light Sources

Parameter	Edge Emitting LED	DFB Laser
Launched power	10–15 dBm	0 dBm
Modulation bandwidth	100–300 MHz	2–10 GHz
Linewidth	100–150 MHz	5–20 MHz
Maximum operating temperature	70°C	50°C
Wavelength stability	0.–0.5 nm/°C	0.–0.5 nm/°C
Applications	LANs, short-haul	MANs, WANs, long-haul
Cost	$100–300	$1000–2000

2.3 PHOTODETECTORS

A photodetector converts an optical signal into an electrical signal, which is an integral part of an optical receiver. Importantly, a receiver should be able to detect low level optical signals without introducing many errors. In digital systems, noise gives rise to bit errors and the signal-to-noise ratio should be sufficiently large to yield an acceptable BER. Four types of photodetectors are available for the design of optical receivers: (1) the avalanche photodiode (APD); (2) the photoconductors; (3) the metal-semiconductor-metal photodiode; and (4) the p-i-n (PIN) diode. The first three types have internal gain, whereas the p-i-n diode does not have any internal gain but has a larger bandwidth.

The choice of a photodetector depends on many considerations, including noise performance, quantum efficiency, bandwidth, high sensitivity, speed, low cost, and high reliability. Based on these considerations, the photodetectors, such as APDs and p-i-n diodes, are commonly used in lightwave systems. Even though the photoconductor is a simpler device that exhibits gain, it has low gain-bandwidth product that makes it unsuitable for practical optical fiber systems. On the other hand, the metal-semiconductor-metal photodiode has gain and bandwidth advantages, and it can be monolithically integrated with a preamplifier. This device operation will be discussed in Chapter 7. For short wavelength systems (0.8–0.9 μm) the silicon material is used, whereas germanium or InGaAs material is used for long wavelength (1.3–1.6 μm) systems. The photodetectors for use in lightwave systems have been discussed in detail in review articles [62,103–108]. In the following subsections, we will review the basic concepts, operation, and characteristics of APDs and p-i-n diodes, including recent developments.

2.3.1 Basic Concepts

The basic mechanism of a photodetector involves creating an electron-hole pair from the incident photon that constitutes the photocurrent. A photodetector usually detects light of the wavelength close to the energy band gap of the semiconductor. The efficiency of a photodetector is measured by its quantum efficiency, which is the ratio of the number of electron-hole pairs collected to the number of incident photons, as given by

$$\eta = \frac{I_{PH}/q}{P_i/h\nu} \tag{2.96}$$

where I_{PH} is the average photocurrent generated by the incident optical power P_i. The quantum efficiency depends on the wavelength, the absorption coefficient of the materials, and the thickness of the material. The quantum efficiency varies by 30–95% in practical photodiodes. Another useful parameter to characterize a photodiode's performance is its responsivity. The responsivity of a photodetector is defined as

$$\Re = \frac{I_{PH}}{P_i} = \frac{\eta q}{h\nu} = \frac{\eta \lambda}{1.24} \tag{2.97}$$

where η is the responsivity measured in A/W, and λ is the wavelength in μm. The responsivity of a typical photodetector at different quantum efficiencies is plotted in Figure 2.26.

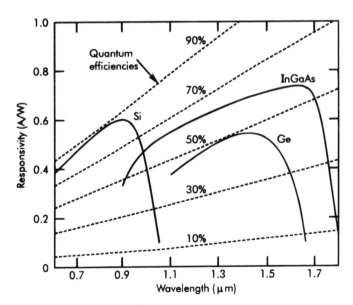

Figure 2.26 Responsivity of a photodetector as a function of wavelength. (*Source:* [12]. Reprinted with the permission of McGraw-Hill.)

2.3.2 p-i-n Photodiodes

The p-i-n diode is a junction diode in which an undoped layer is inserted between the p+ and n+ layers and is operated under reverse bias. The cross-sectional view of the diode along with the energy band diagram, including the carrier generation process, is shown in Figure 2.27. The applied reverse voltage entirely drops across the i-layer because of the low density of free carriers and high resistivity of the layer. In this case, the depletion layer is quite wide, and a small current called dark or reverse saturation current flows through the device. Upon the incidence of photons on the detector, the photons are mainly absorbed in the depletion region, thereby generating the electron-hole pairs. These electron-hole pairs in the depletion region are accelerated in opposite directions by the reverse bias, which gives rise to electrical current. Large numbers of carriers are collected if the depletion layer is wide enough, but the transit time of the diode increases and degrades the high-speed performance. In practical diodes, a tradeoff between the quantum efficiency and bandwidth is necessary. By careful choosing the material properties and device design, very large

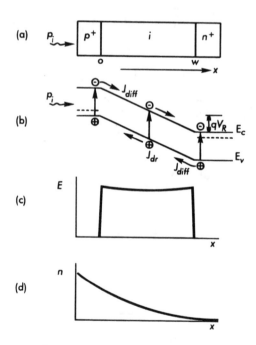

Figure 2.27 Basic operation of a p-i-n diode: (a) p-i-n layer; (b) energy band diagram; (c) electric field; and (d) carrier generation. Here, J_{dr} is the drift current, J_{diff} is the diffusion current, and V_R is the bias voltage. (*Source:* [59]. Reprinted with the permission of Prentice-Hall.)

bandwidths can be obtained. The typical structures of InGaAsP p-i-n diode for long wavelength systems are shown in Figure 2.28 [106].

The back-illuminated mesa structure has an advantage over the top-illuminated planar device structure in that the entire device is photosensitive, which results in lower device capacitance and lower dark current. On the other hand, the planar structure is reliable and is easy to fabricate. The frequency response of the diode depends on the parasitic and junction capacitances of the diode. The frequency response can be obtained from the transfer function of the diode equivalent circuit as [59]

$$H(\omega) = \frac{R_D}{A_1 + j\omega(A_2 - A_3\omega^2) - A_4\omega^2} \tag{2.98}$$

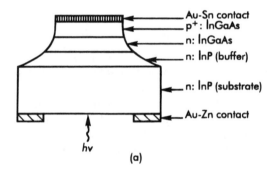

(a)

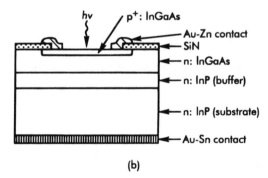

(b)

Figure 2.28 Cross-sectional view of a p-i-n diode structure: (a) back-illuminated mesa structure and (b) top-illuminated planar structure. (*Source:* [7]. Reprinted with the permission of Chapman and Hall.)

where

$$A_1 = R_S + R_L + R_D \tag{2.99a}$$

$$A_2 = R_S R_L C_P + L_S + (R_S + R_L)R_D C_j + R_L R_D C_P \tag{2.99b}$$

$$A_3 = R_S L_S C_P R_D C_j \tag{2.99c}$$

$$A_4 = (R_S R_L C_P + L_S)R_D C_j + R_S L_S C_P + R_D C_P L_S \tag{2.99d}$$

where R_S is the series resistance of the diode, R_L is the load resistance, R_D is the diode shunt or junction resistance, C_j represents the junction capacitance due to the depletion region, C_P is the parasitic capacitance, and L_S denotes the series inductance due to leads. Another important parameter for high-speed optical receivers is the 3-dB bandwidth of the PIN diode, which is given by [59]

$$f_{3\text{ dB}} = \frac{2.8}{2\pi t_r} \tag{2.100}$$

where t_r is the rise time of the impulse response, which is typically about 20 ps, resulting in a 3 dB frequency of about 22 GHz.

Next, let us look at the noise performance of the detector in conjunction with a pre-amplifier. We assume an amplitude modulated optical signal with a unity modulation index, with an r.m.s optical power of $P_i/\sqrt{2}$. The noise performance of a receiver is measured by the signal-to-noise ratio, which is the ratio of mean signal power to mean noise power at load R_L

$$SNR = \frac{i_S^2 R_L}{i_N^2 R_L} \tag{2.101}$$

where i_S^2 denotes the mean square photocurrent, which is given by

$$i_S^2 = \Re^2 \frac{P_i^2}{2} \tag{2.102}$$

and i_N^2 represents the equivalent mean-square noise current, which is given as

$$i_N^2 = [2qB(I_{PH} + I_B + I_D)] + \frac{4k_B T_K B}{R_{eq}} \tag{2.103}$$

where the first term of the equation denotes the shot noise in the device and the second term represents the thermal or Johnson noise. T_K is the temperature in K, and the terms I_B and I_D denote the background and dark currents, respectively. The terms k_B and B denote the Boltzman's constant and bandwidth of the device respectively. The term R_{eq} represents the equivalent resistance, which is given by

$$\frac{1}{R_{eq}} = \left(\frac{1}{R_D} + \frac{1}{R_L} + \frac{1}{R_i}\right) \tag{2.104}$$

where R_i is the amplifier input resistance. Using (2.82) and (2.83) in (2.81) gives

$$SNR = \frac{\dfrac{\Re^2 P_i^2}{2}}{2q(I_{PH} + I_B + I_D)B + \dfrac{4kT_KB}{R_{eq}}} \tag{2.105}$$

From (2.105), it is clear that SNR becomes maximum when the unwanted currents I_B and I_b are minimum and R_{eq} is large.

2.3.3 Avalanche Photodiodes

An avalanche photodiode is similar to a p-i-n diode except that it has internal gain and requires higher reverse voltage for avalanche process or impact ionization. The ionization collision with the lattice can occur, if the photogenerated carriers acquire high enough energy from the applied field. The electric field required to produce such ionization is on the order of 10^4 to 10^5 V/cm. Due to ionization, secondary electron-hole pairs are produced and these electron-hole pairs further drift in the opposite directions along with the primary carrier. This process leads to carrier multiplication and gain. Typical gains on the order of 200 can be achieved in practical APDs. The resulting avalanche process is asymmetric (i.e., the probability of initiating an avalanche is different for holes than it is for electrons). This asymmetry is expressed by the ionization coefficient $k_i = \alpha_e/\alpha_h$, where α_e and α_h are the impact ionization coefficients for electrons and holes, respectively.

The noise performance of an APD is determined by the noise arising from the avalanche process and the shot noise due to the dark current. The noise arising from the avalanche process is random in nature and can be considered a Poisson process [109]. In this Poisson process, there are random fluctuations in the actual distance between the successive ionizing collisions. This gives rise to variations in the total number of carriers generated per primary carrier, which leads to randomness of noise or signal current. The magnitude of the fluctuations or the noise depends on the average avalanche gain. The resulting noise in the avalanche process is characterized by the excess noise factor, which is given by [110]

$$F = M\left[1 - (1 - k_i)\left(\frac{M - 1}{M}\right)^2\right]$$ (2.106)

where M represents the gain of an APD and k_i denotes the ionization constant. The signal-to-noise ratio of an APD is similar to that of a p-i-n diode. In this case, the r.m.s photocurrent is given by

$$i_S^2 = \Re^2 \frac{P_i^2 M^2}{2}$$ (2.107)

and the r.m.s equivalent noise current is given by

$$i_N^2 = 2qB(I_{PH} + I_B + I_D)M^2F + \frac{4k_B T_K B}{R_{eq}}$$ (2.108)

Using (2.107) and (2.108) in (2.101), we get the signal-to-noise ratio as

$$SNR = \frac{\dfrac{\Re^2 P_i^2 M^2}{2}}{2qB(I_{PH} + I_B + I_D)M^2F + \dfrac{4k_B T_K B}{R_{eq}}}$$ (2.109)

The bandwidth of the device is limited by the transit time, which is given by

$$B = \frac{1}{2\pi\tau M}$$ (2.110)

where τ is the transit time, which is usually longer due to the avalanche process. Hence the APD has a smaller bandwidth than the p-i-n diode.

The practical APDs used in optical fiber communications are shown in Figure 2.29. In the case of a Si APD, the incident light is absorbed in the p-type material and hence generates primary carriers for avalanche process. The guard rings at the edge of the p-n junction reduces the excessive leakage currents.

This separate absorption and multiplication (SAM) structure APD [111] yields low leakage currents with a low gain and the dark currents on the order of few pA-nA. The response time of this device is usually low. However, the performance of the device can be improved by inserting a graded band gap quaternary InGaAsP layer between the InP and InGaAs layer, which results in a separate-absorption-graded-multiplication APD or SAGM-APD [112], which is shown in Figure 2.29(b). This device offers better performance characteristics than the SAM-APD diode. Gain

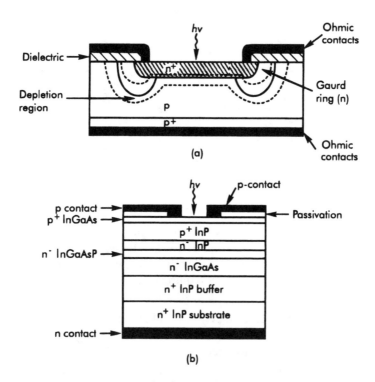

Figure 2.29 APD structures: (a) Si APD and (b) InGaAs SAGM-APD. (*Source:* [59]. Reprinted with the permission of Prentice-Hall.)

bandwidth products of more than 100 GHz can be obtained. Another technique used to improve the response time of the diode involves interposing a thin multiple quantum well structure between the narrow and wideband gap layer [113]. These are called superlattice APDs. Also, Ge APD diodes are available for long-wavelength systems. Nevertheless, Ge APD sensitivities are limited by large dark currents, approximately 10 to 100 times that of InGaAs APDs.

2.3.4 Comparison of Photodetectors

APD structures are more complicated and difficult to fabricate than the p-i-n diodes. Both the diodes require reverse bias voltages, but the APD diodes require higher voltages. In the case of an APD, the avalanche gain is more sensitive to bias and temperature. Despite the APD's disadvantages, the Si APD yields higher sensitivity than the Si-PIN diodes. Currently, SAGM-APD devices offer better performance for long-wavelength systems. Table 2.2 shows the general characteristics of these photodetectors. It can be inferred from Table 2.2 that the best performance in terms

Table 2.2
General Characteristics of Commonly Used Photodetectors

Type of Diode	Wavelength (μm)	Quantum Efficiency	Response Time	Dark Current
Si PIN	0.4–1.1	90%	0.1–5 ns	1–50 nA
Si APD	0.4–1.1	85%	.05–0.3 ns	20–100 nA
Ge APD	0.6–1.6	85%	0.1 ns	20–100 nA
Ge PIN	1–1.6	60%	2.5 ns	100 nA
InGaAs APD	1–1.6	80%	100–200 ps	50 nA
InGaAs PIN	1–1.6	70%	20 ps	10 nA

of bandwidth and sensitivity can be obtained by using a p-i-n diode. However, with proper design, the response speed of an APD should approach that of a p-i-n diode in the near future. Currently, SAGM-APD offers the best performance for practical systems, in spite of higher reverse bias requirements.

2.4 CABLES, SPLICES, CONNECTORS, AND COUPLERS

The practical optical fiber systems require the use of optical cables along with interconnection components such as splices, connectors, and couplers. An optical fiber cable is necessary to protect the fibers from mechanical and environmental factors and to preserve their optical properties. Additionally, the cable should meet several practical requirements, such as ease of installation, and should facilitate fiber splicing and attachment of connectors. A splice is commonly used to permanently join cabled fiber lengths, where the system span is longer than the available cable lengths. The splices are also used for concatenation of shorter fiber sections and for repairing damaged fiber cables. The connector is a detachable component used to connect a transmitter to a cable or a cable to a receiver. The couplers are used to divide or combine or couple the light in a transmission system. The design and fabrication aspects of fiber cables, splices, connectors, and couplers have been discussed in detail in several books [11,13,114] and review articles [115,116]. In the following sections, we briefly review the fiber optic cables, couplers, splices, and connectors used in practical fiber optic systems.

2.4.1 Fiber Optic Cables

A fiber optic cable has a central core element that contains the fibers. The core is surrounded by a plastic jacket such as polyethylene (PE) or polyvinyl chloride (PVC). During the last decade, various types of fiber optic cables have been designed. Some of the commonly used fiber optic cables are: (1) loose tube structure; (2) ribbon cable structure; and (3) slotted core structure. In the case of a loose tube structure, the fibers are loosely packaged in a unit construction and the core is filled typically up to 12 fibers per tube. Multiple tubes are stranded around a center strength member forming the cable core, and the cable is completed with a polyethylene sheath. This structure has advantages of less stress and low microbending loss. The ribbon structure is based on a ribbon that is packaged with 12 fibers between the two adhesive-backed polyester tapes. A maximum of 12 ribbons can be stacked into a rectangular array for fiber counts as high as 144 per cable. A loose plastic tube is extruded over the core, which is filled with a soft water-blocking compound. The filling compound allows free movement of the ribbons within the core. These ribbon cables are compact in size and offer ease of splicing. In the case of a slot structure, the slots or the grooves can follow a helical path or an oscillating path. Up to 10 fibers can be laid in each slot in a slot structure.

2.4.2 Splices

Splicing can be accomplished by using either fusion splices or mechanical splices. The fusion technique is most commonly used for joining multimode as well as single mode fibers. In this method, the fibers are heated locally at the joint until they are softened and can be fused together. Electric arc or gas flame or laser sources can be used as a heat source. The mechanical splices make use of suitable fixtures to obtain good alignment. Fixtures such as capillary splices and groove splices have been developed. The average insertion loss resulting from splicing ranges from 0.1 to 0.2 dB for both multimode and single mode fibers.

2.4.3 Connectors

The design of connectors should take into account low insertion loss, ease of field assembly, environmental stability, and reliability. The two most popular connectors are: (1) the butt-joint structure and (2) the expanded-beam structure. The average insertion loss of these connectors ranges from 0.2 to 0.3 dB.

2.4.4 Couplers

In optical fiber systems such as LANs or point-to-point links, the signal has to be distributed to many subscribers, which require couplers. The couplers can be classified into different types, based on their functions. Some of these couplers are shown in Figure 2.30. The function of an optical splitter is to divide the optical signal carried by a single fiber into two signals as shown in Figure 2.30(a), whereas the optical combiner combines the signals from two fibers into one as shown in Figure 2.30(b). The couplers having input or output ports in excess of two are usually called $N \times N$ star or $1 \times N$ tree couplers and are depicted in Figures 2.30(c,d). These are used in

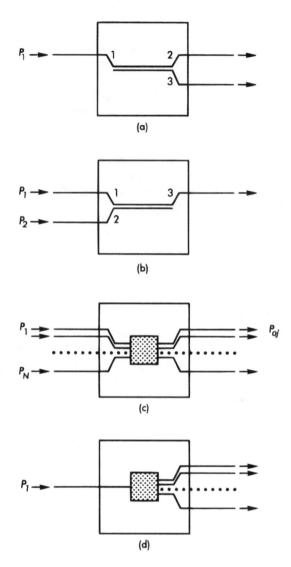

Figure 2.30 Couplers: (a) optical splitter; (b) optical combiner; (c) star coupler; and (d) tree coupler.

LANs, which are supposed to produce uniform division of the input power at the output ports. The optical couplers are made of multimode or singlemode fibers or by using miniaturized devices such as graded-refractive-index rod lenses.

To understand functional characteristics of a coupler, let us consider the schematic representation of an $N \times N$ star coupler, which is shown in Figure 2.30(c). The output power at the jth output port is given by

$$P_{oj} = \sum_i \alpha_{ji} P_i \qquad (2.111)$$

where α_{ji} represents the coupling coefficient between the ith input port and the jth output port, which takes into account the branching effect of the coupler and the loss caused by reflections, scattering, and various imperfections. The insertion loss of the coupler (decibels) along the ith to the jth port is given by

$$L_{ij} = -10 \log(\alpha_{ji}) \qquad (2.112)$$

For example, a typical 32-by-32 single-mode star coupler will have an insertion loss of 1.0–1.5 dB. Another parameter of a coupler is the directivity, which is measured by the isolation between its input ports. When power is launched into the ith port, the isolation between the ith and kth ports is defined (in decibels) as

$$D_{ki} = -10 \log\left(\frac{P_r}{P_i}\right) \qquad (2.113)$$

where P_r is the amount of undesired power at the kth port. An isolation of typically -40 to -50 dB is usually required. The $N \times N$ star coupler can be built by using only 2×2 couplers or in combination with combiners and splitters. In general, the required number of 2×2 couplers to build an $N \times N$ coupler is given by

$$N_C = \frac{N}{2} \log_2 N \qquad (2.114)$$

where N_C denotes the required number of 2×2 couplers. For example, the number of couplers required to build a 16×16 star coupler is 32. Another way of building a 16×16 coupler is to use a 2×2 star coupler in combination with four 4×1 combiners at the input and four 1×4 splitters at the output.

2.5 SOURCE-FIBER COUPLING

Lightwave communication systems require efficient launching of light from a light source into a fiber, so that the repeater spacing is not severely limited. We have learned that when the light rays impinge on the fiber core at an angle less than the acceptance angle, the total internal reflection occurs and the rays are guided along the fiber. Therefore, the fraction of light (or coupling efficiency) mainly depends on the angular distribution of the light. Other considerations such as numerical aperture, core size, refractive index profile, relative index difference, and radiance should be taken into account.

2.5.1 LED-Fiber Coupling

The maximum power that can be coupled from an LED into a fiber is given by

$$P_{max} = I_0 A_s \theta_C \cong I_0 A_s \pi (NA)^2 \qquad (2.115)$$

where I_0 is the radiance (P_{out}/emitting area) and θ_C is the acceptance angle, A_s denotes the smaller value of the areas of the emitter and the fiber core, and NA represents the numerical aperture. The coupling efficiency is defined as

$$\eta_C = \frac{P_{max}}{P_{in}} \qquad (2.116)$$

where P_{in} denotes the optical power generated by an LED. The maximum efficiency can be obtained if the product $I_0 A_s$ is maximized in (2.115). The two most popular coupling methods are: (1) butt coupling and (2) lens coupling. Butt coupling is the simplest method used to couple the LED (particularly the surface emitting structure) into a fiber. In this approach, the LED surface emitting area should not be larger than the fiber core area for maximum power coupling. The butt-coupling efficiency based on the Lambertian intensity profile can be obtained by integrating the radiation intensity over the LED emitting area and the acceptance angle. The coupling efficiency is therefore given by

$$\eta_C = T_R (NA)^2 \qquad (2.117)$$

for step-index fiber and

$$\eta_C = T_R (NA)^2 \left(1 - \frac{1}{2D_{ce}^2}\right) \qquad (2.118)$$

for graded-index fiber where T_R denotes the transmittivity and D_{ce} is the ratio of diameters of the fiber core to the emitting surface. From the above equations, it can be inferred that the coupling efficiency of a graded-index fiber approaches that of a step-index fiber when the fiber core diameter is much larger than the emitting surface diameter. A coupling efficiency of about 26–30% can be obtained by the butt-coupling method.

The coupling efficiency can be improved further by using the lens coupling method. In this method, lenses are used to collimate the emission for the LED. Several lens coupling schemes have been studied in the literature [117,118]. All of these schemes are aimed at magnifying the emission area so that the size of the image of the LED emitting area matches the fiber core for achieving maximum efficiency. In general, lens coupling is more advantageous than butt coupling when the fiber core diameter is much larger than the source diameter. In the case of edge emitting LED coupling, lens coupling schemes are more efficient than the butt-coupling methods.

2.5.2 Laser Diode—Fiber Coupling

The radiated light from laser sources such as index guided structures can be approximated by a Gaussian profile. The optical coupling between a laser source and a fiber requires a matching of the Gaussian beam waists of the laser and that of a fiber core. Index guide structures have a waist radius of 1.5 μm and the waist for SMF is about 4.5 μm. In the past, several methods have been reported in the literature [119]. These methods can be broadly classified into two types: (1) single-lense schemes and (2) expanded-beam schemes.

The expanded-beam design uses lenses to create enlarged images of the beam waists of both the laser and the fiber. The alignment accuracy is the same for both methods, but the expanded-beam method allows the insertion of an isolator between the lenses to minimize any reflections. Coupling efficiencies of about 50% can be obtained with these schemes.

2.6 SOURCE-DRIVER CIRCUITS

Source-driver circuits provide the electrical power for the light sources and modulate the light output in accordance with the information-bearing signal. The basic requirements for a driver circuit shall include the maximum drive current capability, high-speed switching response, temperature control, and reliability.

For LEDs, simple common emitter transistor switches can be utilized to build the driver circuits, whereas in practical laser driver circuits, care should be given for temperature control and the effects of aging. Particularly, in InGaAsP lasers, the threshold and external quantum efficiency are a function of temperature changes and hence separate circuits must be designed to monitor and regulate the operating temperature. Also, the driver circuit should have a feedback mechanism to adjust the current. In recent years, monolithic integration of light sources and electronic driver circuits has been proposed, and it offers several advantages for high-speed operation and ease of packaging [120,121].

2.7 SYSTEM DESIGN

The basic characteristics of the individual building blocks of an optical fiber communication system were discussed in the previous sections. In this section, we assemble all the necessary individual blocks to form a system. Even though different system link architectures are possible for a variety of applications, we shall examine a simple point-to-point link architecture to understand the system design concepts.

The design of a link involves many interrelated system requirements and parameters. The key requirements are: (1) a *BL* product and (2) a *BER* for a digital link or SNR for an analog link. To meet these requirements, depending on the application, the system designer has to consider the following eight parameters: (1) the wavelength window; (2) system component considerations; (3) performance degradation parameters; (4) the modulation format; (5) system cost; (6) reliability; (7) lifetime; and (8) upgradability. Several techniques, such as link power budget, rise time budget, and computer models, can be utilized to design a system link.

In this section, we use power budget as a basis for the system design. The optical power received at the receiver mainly depends on the amount of light launched into the fiber and on the losses incurred in a link due to the fiber, splices, and connectors. For a point-to-point link, considering the transmitter and receiver power levels, as well as the losses, we have

$$(P_T - P_R) - (\alpha_f + \alpha_S + \alpha_C + M_S)L \geq 0 \tag{2.119}$$

where the first term represents the system gain and the second term denotes the system losses. From (2.119), the maximum transmission distance between the optoelectronic regenerators is given by

$$L = \frac{(P_T - P_R)}{(\alpha_f + \alpha_S + \alpha_C + M_S)} \tag{2.120}$$

where P_T is the launched power into the fiber; P_R is the receiver sensitivity; α_f represents the fiber loss; the splice loss is denoted by α_S; α_C is the connector loss; and M_S denotes the system margin, which is usually about 6–8 dB. The losses are measured in terms of decibels, whereas the transmitter and receiver power levels are measured in terms of dBm. Equation (2.120) can be used to design a point-point link for the given system requirements and parameters.

Problems

The following problems relate to Section 2.1.

Problem 2.1

Assume a step-index fiber having a core refractive index $n_1 = 1.46$ and the refractive index of the cladding $n_2 = 1.40$. Calculate the following: (a) the critical angle for total reflections; (b) the maximum acceptance angle for the fiber, assuming the outer medium as air; (c) the numerical aperture; and (d) relative core-cladding difference.

Problem 2.2

Assume a 1.3 μm step index fiber with n_1 = 1.468 and n_2 = 1.460 and a core radius of 25 μm. Sketch the eigen value equations (2.33) for m = 0, 1, and 2, using a table of Bessel functions. Also, find the propagation constants for the lowest and highest order modes for m = 0.

Problem 2.3

For the fiber specified in problem 2.2, obtain the complete equations for E_z, H_z for the region inside the core as well as for the region outside the core. Also, find the number of propagating modes.

Problem 2.4

Determine the propagating modes of a step index fiber with V = 5.8.

Problem 2.5

Design a single mode fiber for a cutoff wavelength 1.55 μm, assuming a step-index profile having refractive index of the core as n_1 = 1.48 and n_2 = 1.455. Also, find the numerical aperture and the maximum acceptance angle.

Problem 2.6

Sketch the refractive index profiles of a graded-index fiber for s = 1, 2, 3. Assume a core radius of 20 μm with graded-index profile having n_1 = 1.48 and Δ = 0.01. Also, calculate the number of propagating modes.

Problem 2.7

For the step-index fiber specified in problem 2.1, calculate the BL product. Also, what is the maximum operating bit rate for the system if the repeater spacing is limited to 15 km

Problem 2.8

Calculate the BL product of a graded index fiber specified in problem 2.5. What is the maximum operating bit rate if the repeater spacing is limited to 30 km.

Problem 2.9

Assume a single mode fiber with $n_1 = 1.48$, $n_2 = 1.466$, and a core radius of 20 μm operating at a wavelength of 1.55 μm. Calculate the BL product of the fiber. What is the maximum operating bit rate of the fiber, if the repeater spacing is limited to 100 km.

Problem 2.10

Calculate the critical power required for a single channel system with a repeater spacing of 100 km assuming SRS and SBS with $\alpha = 0.3$ dB/km, $g_0 = 7 \times 10^{-9}$ cm/W, and $A_e = 90$ μm^2. Assume a conventional single mode fiber and $\Delta\nu_P/\Delta\nu_B \ll 1$.

Problem 2.11

Calculate the power per channel required for a 1 dB system penalty, assuming SRS with 100 channels spaced at 5 GHz.

Problem 2.12

Calculate the number of channels that can be transmitted over a single mode fiber considering the effect of CPM. Assume the r.m.s. power fluctuation of 0.2 mW and the r.m.s. phase fluctuations of 0.2 radians. Also, calculate the maximum power per channel for a 1 dB system penalty.

Problem 2.13

Find the frequency components generated due to FPM, if three co-propagating waves propagate in the fiber. Also, sketch the resulting spectrum.

The following problems relate to Section 2.2.

Problem 2.14

Find the emission wavelength of $Al_{0.05}$ $Ga_{0.95}$ As and $In_{0.7}$ $Ga_{0.3}$ $As_{0.6}$ $P_{0.4}$ materials.

Problem 2.15

Calculate the wavelength of the photons absorbed or radiated for a two-level system having $E_c = 1.4$ eV and $E_v = 0$ eV.

Problem 2.16

Assume a Fabry-Perot laser diode emitting light at 0.8 μm having a cavity length of 400 μm with an internal cavity losses of 15/cm. The mirror facet reflectivities are 0.32 and 0.35 respectively. (a) Calculate the optical gain as well as the threshold currents at lasing. (b) Find the resulting power output from the laser, if the laser is biased at 50 mA, assuming the internal quantum efficiency of 60%.

Problem 2.17

Assume a 1.55 μm semiconductor laser having a cavity length of 250 μm with refractive index of 3.4 in the active layer. Calculate the spacing between various oscillating modes.

Problem 2.18

Consider the PI characteristics of a semiconductor laser. The laser power output is 2.5 mW for a diode current of 40 mA. The power output is increased to 5 mW when the current is 45 mA. Find (a) the threshold current and (b) the maximum and minimum output power levels, when the diode current is modulated with a square wave, with I_{avg} = 40 mA, I_{min} = 30 mA, and I_{max} = 55 mA.

Problem 2.19

Consider the rate equations given in (2.78) and (2.79). Derive the following steady state solutions for the carrier and photon densities.

$$n_0 = \frac{S(Cn_m + 1/\tau_p)}{(CS + \Gamma/\tau)}$$

$$S_0 = \frac{\tau_p}{qd}(J_0 - J_{th})$$

Hint: Set the derivatives in (2.78) and (2.79) to zero and then solve for n and S.

Problem 2.20

Derive (2.88). Assume $\Gamma = 0$ and use the following approximation for the term involving

$$[n(t) - n_m]S \approx (n_0 - n_m)S + [(n_0 - n_m)S + Sn_1]e^{j\omega t}$$

Problem 2.21

Assume a pulse modulation on a laser diode. The laser diode is biased at 30 mA and assume a laser threshold current of 20 mA. Calculate the delay time of the optical pulse, if the radiative and nonradiative carrier lifetimes are 10 ns and 100 ns respectively. Also, calculate the resulting resonant frequency for a photon lifetime of 10 ns.

Problem 2.22

Assume a 0.8 μm LED, which is modulated with a current pulse of 20 mA for a binary bit 1 and 0 mA for a bit 0. If the quantum efficiency of the LED is 0.2, what is the output power of the LED, when a binary bit 1 is transmitted?

Problem 2.23

Plot the frequency response of an LED using (2.94) for $\tau = 10$ ns. Also, find the 3 dB bandwidth of the LED. Assume $\tau_r = \tau$.

The following problems relate to Section 2.3.

Problem 2.24

Calculate the quantum efficiency of a photodetector at a wavelength of 1.3 μm, if the incident optical power of 1W generates an average photocurrent of 0.8 A. Also, find the responsivity of the diode.

Problem 2.25

Plot the magnitude of the frequency response of the InGaAs PIN diode, given the following equivalent circuit parameters: $R_D = 100$ M Ohms, $R_L = 100$ Ohms,

R_S = 10 Ohms, C_j = 100 pF, C_P = 10 pF, and L_S = 50 pH. From the response, calculate the 3 dB bandwidth of the detector.

Problem 2.26

It is often convenient to use the NEP (noise equivalent power) parameter to measure the minimum detectable power of a photodetector. NEP is calculated from a unity SNR with a 1 Hz bandwidth. Derive the expression for NEP for a PIN and APD diodes using the corresponding SNR equations.

Problem 2.27

Using the expression derived in Problem 2.26, Calculate the value of NEP for a InGaAs PIN diode at 1.5 μm given the following parameters: R_i = 10 M Ohms, R_D = 100 M Ohms, R_L = 10 M Ohms, I_B = 1 pA, and I_D = 1 nA. Assume the quantum efficiency of the diode as 70%.

Problem 2.28

Assume an APD diode with a responsivity of 0.8 A/W at 1.55 μm with a gain of 100. If 10^{12} photons/sec are incident on the device, find the quantum efficiency and the resulting photocurrent.

Problem 2.29

Assume an APD operating at a wavelength of 0.85 μm with the incident signal power of 10 nW. Calculate the SNR of the device for the given parameters: Q = 0.85, M = 200, I_B = I_D = 10 nA, R_{eq} = 1 k Ohms, k_i = 0.5, and B = 1 MHz.

The following problems relate to Section 2.5.

Problem 2.30

An LED with a circular emitting area of diameter 50 μm has a lambertian emission profile with a radiance 150 W/cm^2 at 85 mA. What is the maximum power that can be coupled into a step-index fiber which has a diameter of 100 μm with $NA = 0.2$. Also, calculate the corresponding coupling efficiency if the transmittivity is 0.6.

The following problems relate to Section 2.7.

Problem 2.31

Find the maximum transmission distance of the system operating at 0.85 μm assuming the following parameters: (a) laser diode-fiber coupled power of 0 dBm; (b) step-index fiber loss of 3.0 dB/km; (c) PIN diode receiver sensitivity of −60 dBm; (d) connector and splice losses of 2.5 dB. Assume a 6-dB system margin.

Problem 2.32

Design a 1.3 μm system for a bit rate of 1 Gb/s and repeater spacing of 100 km based on the given parameters: (a) laser diode-fiber coupled power of 0 dBm, (b) fiber loss of 3 dB/km, (c) splice and connector losses of 2 dB, (d) receiver sensitivity of −55 dBm, and (e) system margin of 8 dB. Choose the appropriate transmitter and receiver components.

References

[1]   Kao, C. K., and G. A. Hockman, "Dielectric-fiber surface waveguides for optical frequencies," *Proc. IEE*, Vol. 113, 1966, pp. 1151–1158.
[2]   Kapron, F. P., D. B. Keck, and R. D. Maurer, "Radiation losses in glass optical waveguides," *Appl. Phys. Lett.*, Vol. 17, 1970, pp. 423–425.
[3]   Marcuse, D., *Light Transmission Optics*, Princeton, New Jersey: Van Nostrand Reinhold, 1972.
[4]   Kapron, F. P. and Keck, D. B., "Pulse transmission through a dielectric optical waveguide," *Appl. Optics*, Vol. 10, No. 7, pp. 1520–1523, 1971.
[5]   Miller S. E., ed. *Optical Fiber Telecommunications*, New York, Academic Press, 1979.
[6]   Midwinter J. E., *Optical Fibers for Transmission*, New York, John Wiley, 1979.
[7]   Lin, C., *Optoelectronic Technology and Lightwave Communications Systems*, Chapman and Hall, New York, 1989.
[8]   Chen, C., *Elements of Optoelectronics and Fiber Optics*, Irwin, Chicago, 1996.
[9]   Snyder A. W. and Love J. D., *Optical Waveguides*, Chapman and Hall, London, 1984.
[10] Neumann, *Single-mode Fibers: Fundamentals*, Springer Verlag, Berlin, 1988.
[11] Muratam H., *Handbook of Optical Fibers and Cables*, Marcel Dekker, New York, 1988.
[12] Keiser, G., *Optical Fiber Communications*, New York: McGraw-Hill; 2nd ed., 1991.
[13] Miller S. E, and I. P. Kaminow, *Optical Fiber Telecommunications II*, Academic Press, New York, 1988.
[14] Miller S. E., Marcatili, E. A. J., and T. Li, "Research toward optical fiber transmission systems, Part I: the transmission medium, *Proc. IEEE*, Vol. 61, 1973, pp. 1703–1726.

[15] Gloge, D. "Propagation effects in optical fibers," *IEEE Trans. Microwave Theory Tech.*, Vol. MTT-23, 1975, pp. 106–120.

[16] Kapron F. P., "Maximum information capacity of fiber optic waveguides," *Electronics Letters*, Vol. 13, 1977, pp. 96–97.

[17] Li T., "Structures, Parameters, and Transmission Properties of Optical Fibers," *Proc. of IEEE*, Vol. 68, 1980, pp. 1175–1180.

[18] Marcuse D. and C. Lin, "Low dispersion single mode fiber transmission-The questions of practical versus theoretical bandwidth," *IEEE J. Quantum Electronics*, Vol. QE-17, 1980, pp. 869–878.

[19] Inada, K., "Recent progress in fiber fabrication techniques by vapor-phase axial deposition," *IEEE J. Quantum Electron.*, Vol. QE-18, 1982, pp. 1424–1432.

[20] Saifi, M. A, S. J. Tang, L. G. Cohen, and J. Stone, "Triangular profile single mode fiber," *Opt. Letts.*, Vol. 7, 1982, pp. 43–45.

[21] Bhagavatula, V. A., J. E. Ritter, and R. A. Modavis, "Bend optimized dispersion shifted single mode designs," *IEEE Journal of Lightwave Technology.*, Vol. 3, 1985, pp. 954–957.

[22] Ohasi, M. N., et al., "Characteristics of bend-optimized convex dispersion shifted fiber," *First Optoelectronics Conf., OEC" 86*, Tokyo, 1986, PDP-22.

[23] Bachmann, P. K., D. Leers, and D. U. Wiechert, "The bending performance of matched cladding, depressed-cladding and dispersion flattened single mode fibers," *OFC/IOOC'87*, 1987, PDP1.

[24] Gloge, D., "Weakly guiding fibers," *App. Opt.*, Vol. 10, 1971, pp. 2252–2258.

[25] Ankiewicz, A. and G. Peng, "Generalized Gaussian approximation for single mode fibers," *IEEE Journal of Lightwave Technology*, Vol. 10, No. 1, 1992, pp. 22–27.

[26] Marcuse, D., "Gaussian approximation of the fundamental modes of graded-index fibers," *J. Opt. Soc. Amer.*, Vol. 68, 1978, pp. 103–109.

[27] Henry, P., "Lightwave Primer," *IEEE J. Quantum Electronics*, Vol. QE-21, No. 12, 1985, pp. 1862–1877.

[28] White, K. I. and B. P. Nelson, "Zero total dispersion in step index monomode fibers at 1.3 and 1.55 μm," *Electronics Letters*, Vol. 15, 1979, pp. 396–397.

[29] Jaunart, E., P. Crahay, P, et al. "Chromatic dispersion modeling of single mode optical fibers: A detailed analysis," *IEEE Journal of Lightwave Technology*, Vol. 12, No. 11, 1994, pp. 1910–1915.

[30] Sarma, A., et al., "Chromatic dispersion in single mode fiber with arbitrary index profiles," *IEEE Journal of Lightwave Technology*, Vol. 7, 1989, pp. 1919–1923.

[31] Mammel, W. L., and L. G. Cohen, "Numerical prediction of fiber transmission characteristics from arbitrary refractive-index profiles," *Appl. Opt.*, Vol. 21, 1982, pp. 699–703.

[32] Noda J., K. Okamoto, and Y. Sasaki, "Polarization maintaining fibers and their applications," *IEEE Journal of Lightwave Technology*, Vol. LT-4, 1986, pp. 1071–1089.

[33] Payne, D. N., A. J. Barlow, and J. Hansen, "Development of low and high birefringence optical fibers," *IEEE J. Quantum Electronics*, Vol. 18, 1982, pp. 477–488.

[34] Okamoto, K T. Hosaka, and Y. Sasaki, "Linearly single polarization fibers with zero polarization mode dispersion," *IEEE J. Quantum Electronics*, Vol. QE-18, 1982, pp. 496–503.

[35] Tajima, K, M. Ohashi, and Y. Sasaki, "A new single polarization optical fiber," *IEEE Journal of Lightwave Technology*, Vol. 7, 1989, pp. 1499–1503.

[36] Hall D. W., Ed., "Special mini issue on dispersion compensation," in *IEEE Journal of Lightwave Technology*, Vol. LT-12, No. 10, 1994, pp. 1706–1759.

[37] Gautheron, O., G. Grandpierre, L. Pierre, J. P. Thiery, and P. Kretzmeyer, "252 km repeaterless 10 Gb/s transmission demonstration," in *Tech. Dig. OFC/IOOC'93*, San Jose, CA, 1993, paper PD11.

[38] Dugan J. M., A. J. Price, M. Ramadan, D. L. Wolf, E. F. Murphy et al, "All-Optical fiber based 1550 nm dispersion compensation in a 10 Gb/s, 150 km transmission experiment over 1310 nm optimized fiber," in *Proc. OFC "92*, San Jose, 1992, Paper PD14.

[39] Antos A. J. and D. K. Smith, "Design and characterization of dispersion compensating fiber based on the LP mode," *IEEE Journal of Lightwave Technology*, Vol. LT-12, No. 10, 1994, pp. 1739–1745.

[40] Wedding, B., B. Franz, and B. Junginger, "10 Gb/s optical transmission up to 253 km via standard single mode fiber using the method of, dispersion supported transmission," *IEEE Journal of Lightwave Technology*, Vol. 12, No. 10, pp. 1720–1727, 1994.

[41] Chraplyvy, A. R., "Limitations on lightwave communications imposed by optical fiber nonlinearites," *IEEE Journal of Lightwave Technology*, Vol. 8, No. 10, 1990, pp. 1548–1557.

[42] Chraplyvy, A. R. and P. S. Henry, "Performance degradation due to stimulated Raman scattering in wavelength division multiplexed optical fiber systems," *Electronics Letters*, Vol. 19, 1983, p. 641.

[43] X. P. Mao, et al., "Stimulated Brillouin threshold dependence on fiber type and uniformity," *IEEE Photonics Technology Letters*, Vol. 4, No. 1, 1992, pp. 66–68.

[44] Lichtman, E., "Performance degradation due to four-wave mixing in multichannel coherent optical communication systems," *Journal of Optical Communications*, Vol. 12, No. 2, 1991, pp. 53–58.

[45] Inoue, K. and H. Toba, "Influence of fiber four-wave mixing on multichannel direct detect-ion transmission systems," *IEEE Journal of Lightwave Technology*, Vol. 10, No. 3, 1992, pp. 350–360.

[46] Inoue, K. and H. Toba, "Theoretical evaluation of error rate degradation due to fiber-four wave mixing in multichannel FSK envelope detection transmissions," *IEEE Photonics Technology Letters*, Vol. 10, No. 3, 1992, pp. 361–363.

[47] Inoue, K. et al., "Crosstalk and power penalty due to fiber four-wave mixing in multichannel transmissions," *IEEE Journal of Lightwave Technology*, Vol. 12, No. 8, 1994, pp. 1423–1429.

[48] Zou, X. Y. et al., "limitations in 10 Gb/s WDM optical fiber transmission when using a variety of fiber types to manage dispersion and nonlinearities," *IEEE Journal of Lightwave Technology*, Vol. 14, No. 6, 1996, pp. 1144–1152.

[49] Inoue, K., "Reduction of fiber four-wave mixing influence using frequency modulation in multichan-nel IM/DD transmission," *IEEE Photonics Technology Letters*, Vol. 4, No. 11, 1992, pp. 1301–1303.

[50] Inoue, K., "Suppression technique for fiber four-wave mixing using optical multi-/demultiplexers and a delay line," *IEEE Journal of Lightwave Technology*, Vol. 11, No. 3, 1993, pp. 455–461.

[51] Tkach, R. W., A. R. Chraplyvy, F. Forghieri, A. H. Gnauck, and R. M. Derosier, "Four-Photon mixing and high-speed WDM systems," *IEEE Journal of Lightwave Technology*, Vol. 13, No. 5, 1995, pp. 841–849.

[52] Fishman, D. A., and J. A. Nagel, "Degradations due to stimulated scattering in multigigabit intensity modulated fiber-optic systems," *IEEE Journal of Lightwave Technology*, Vol. 11, No. 11, 1993, pp. 1721–1728.

[53] Hall, R. N., Fenner, G. E., Kingsley, J. D. Soltys, T. J., and Carlson, R. O., "Coherent light emission from GaAs junctions," *Apply. Phys., Lett.*, Vol. 9, p. 366, 1962.

[54] Nathan, M. I., Dumke, W. P., Burns, G., Dill, F. H., Jr, and Lasher, G., "Stimulated emission of radiation from GaAs p-n junctions," *Apply. Phys., Lett.*, Vol. 1, p. 62, 1962.

[55] Kressel, H., and J. K. Butler, *Semiconductor Lasers and Heterojunction LED's*, Academic Press, New York, 1977.

[56] Casey, H. C. Jr. and M. B. Panish, *Heterostructure Lasers: Part A, Fundamental Principles, Part B, Materials and Operating Characteristics*, Academic Press, New York, 1978.

[57] Saleh, B. E. A. and M. C. Teich, *Fundamentals of Photonics*, John Wiley & Sons Inc., New York, 1991.

[58] Agrawal, G. P. and N. K. Dutta, *Long-wavelength Semiconductor Lasers*, Van Nostrand Reinhold, New York, 1986.

[59] Bhattacharya, P. B., *Semiconductor Optoelectronic Devices*, Prentice-Hall, New Jersey, 1994.

[60] Botez, D., and G. J. Herskowitz, "Components for optical communication systems: A Review," *Proc. IEEE*, Vol. 68, No. 6, 1980, pp. 689–731.

[61] Kressel H. Ed., "Semiconductor devices for optical communication," in *Topics in Applied Physics*," Vol. 39, Chap. 2, Springer-Verlag, 1980.

[62] Newman D. H., and Ritchie S., "Sources and detectors for optical fiber communications applications: the first 20 years," *IEE Proceedings*, Vol. 133, Pt. J, No. 3, 1986. pp. 213–229.

[63] Nakamura, M. and S. Tsuji, "Single mode semiconductor injection lasers," *IEEE J. of Quantum Electronics*, Vol. QE-17, 1981, pp. 994–1005.

[64] Channin D. J., "Emitters for fiberoptic communications," *Laser Focus*, Vol. 18, No. 11, 1982, pp. 105–113.

[65] Special issue on semiconductor lasers, IEEE J. of Quantum Electron., Vol. QE-25, no. 6, 1989.

[66] Nahory, R. E., M. A. Pollack, W. D. Johnston, and R. L. Barnes, "Bandgap versus composition and demonstration of vegard's law for $In_{1-x}Ga_xAs_xP_{1-y}$ lattice matched to InP," *Appl. Phys. Lett.*, Vol. 33, 1978, pp. 659–661.

[67] Kirkby, P. A., A. R. Goodwin, G. H. B. Thompson, and P. R. Selway, "Observations of self-focusing in stripe geometry semiconductor lasers and the development of a comprehensive model of their operation," *IEEE J. Quantum Electron.*, Vol. QE-13, 1977, pp. 703–719.

[68] Turley, S. E. H., G. D. Henshall, P. D. Greene, V. P. Knight, D. M. Moule, and S. A. Wheeler, "Properties of inverted rib-waveguide lasers operating at 1.3 μm wavelength," *Electronics Letters*, Vol. 17, 1981, pp. 868–870.

[69] Armistead, C. J., S. A. Wheeler, R. G. Plumb, and R. W. Musk, "Low threshold ridge waveguide lasers at 1.5 μm," *Electronics Letters*, Vol. 22, 1986, pp. 1145–1147.

[70] Hirao, M. A. Dori, S. Tsuji, M. Nakamura, and K. Aiki, "Fabrication and characterization of narrow stripe InGaAsP/InP buried heterostructure lasers," *J. Appl. Phys.*, Vol. 51, 1980, pp. 4539–4540.

[71] Nelson, R. J., R. B. Wilson, P. D. Wright et. al, "CW electrooptical properties of InGaAsP buried-heterostructure lasers," *IEEE J. of Quantum Electronics*, Vol. QE-17, 1981, pp. 202–207.

[72] Vodhanel, R. S., A. F. Elrefaie, M. Z. Iqbal, R. E. Wagner, J. L. Gimlett, and S. Tsuji, "Performance of directly modulated DFB lasers in 10-Gb/s ASK, FSK, and DPSK lightwave systems," *IEEE Journal of Lightwave Technology*, Vol. 8, No. 9, 1990, pp. 1379–1386.

[73] Kogelink, H. and C. V. Shank, "Coupled-wave theory of distributed feedback lasers," *J. Appl. Phys.*, Vol. 43, 1972, pp. 2327–2335.

[74] Streifer, W., R. D. Burnham, and D. R. Scifres, "Effect of external reflectors on longitudinal modes of distributed feedback lasers," *IEEE J. Quantum Electron.*, Vol. QE-11, 1975, pp. 154–161.

[75] Zah, C. E., et al., "Performance of 1.5 μm quarter-wavelength shifted DFB-SIPBH laser diodes with electron-beam defined and reactive ion etched gratings," *OFC'89*, Houston, TX, 1989, Paper WB5.

[76] Tsang, W. T, ed., "The cleaved-coupled cavity laser," in *Semiconductor and Semimetals: Lightwave Communication Technology*, Academic Press, Vol. 22B, London, 1985, pp. 257–373.

[77] Liou, K. Y. et al. "Single-longitudinal mode stabilized graded-index rod external coupled-cavity lasers," *Appl. Phys. Lett.*, Vol. 45, 1984, pp. 729–931.

[78] Tsang, W. T. ed. "Quantum confinement heterostructure semiconductor lasers," in *Semiconductor and Semimetals*, Chapter 7, Academic press, New-York, 1987.

[79] Yariv, A., "Quantum well semiconductor lasers are taking over," *IEEE Circuits and Devices Mag.*, Vol. 5, No. 6, 1989, pp. 25–27.

[80] Special issue on semiconductor quantum well heterostructures and superlattices, *IEEE J. of Quantum Electron.*, Vol. QE-24, no. 8, 1988.

[81] Zory, P. S. *Quantum Well Lasers*, New York, Academic Press, 1993.

[82] Joyce, W. B. "Carrier transport in double heterostructure active lasers," *J. Appl. Phys.*, Vol. 53, No. 11, 1982, pp. 7235–7238.

[83] Papannareddy, R., W. E. Ferguson, and J. K. Butler, "Four models of lateral current spreading in double-heterostructure stripe-geometry lasers," *IEEE J. Quantum Electron.*, Vol. QE-24, No. 1, 1988, pp. 60–65.

[84] Shore, K. A, T. E. Rozzi, and H. H. Veld, "Semiconductor laser analysis: general method for characterizing devices of various cross-sectional geometries," *IEE Proc.* Vol. 127, Part I, No. 5, 1980, pp. 221–229.

[85] Reinhart, F. K., L. Hayashi, and M. B. Panish, "Mode reflectivity and waveguide properties of double-heterojunction injection lasers," *J. Appl. Phys.*, Vol. 42, 1971, pp. 4466–4479.

[86] Butler, J. K. "A rigorous boundary value solution for the lateral modes of stripe geometry injection lasers," *IEEE J. Quantum Electron.*, Vol. QE-14, No. 7, 1978, pp. 507–513.

[87] Meissner P., E. Patzak, and D. Yevick, "A self-consistent model of stripe-geometry lasers based on beam propagation method," *IEEE Quantum Electron.*, Vol. QE-20, No. 8, 1984, pp. 899–905.

[88] Joyce, W. B., and R. W. Dixon, "Thermal resistance of heterostructure lasers," *J. Appl. Phys.*, Vol. 46, No. 2, 1975, pp. 855–862.

[89] Nakwaski, W., "Dynamic thermal properties of stripe-geometry laser diodes," *IEE Proc.*, Vol. 131, Part I, No. 3, 1984, pp. 94–102.

[90] Papannareddy, R., W. E. Ferguson, and J. K. Butler, "A generalized thermal model for stripe-geometry injection lasers," *J. Appl. Phys.*, Vol. 62, No. 9, 1987, pp. 3565–3569.

[91] Ettenberg, M., C. J. Nuese, and H. Kressel, "The temperature dependence of threshold current for double heterojunction lasers," *J. Appl. Phys.*, Vol. 50, No. 4, 1979, pp. 2949–2950.

[92] Boers, P. M, M. T. Vlaardingerbroek, and M. Danielsen, "Dynamic behavior of semiconductor lasers," *Electronics Letters*, Vo. 11, 1975, pp. 128–132.

[93] Paoli, T. L. and J. E. Ripper, "Direct modulation of semiconductor lasers," *Proc. IEEE*, Vol. 58, 1970, pp. 1457–1565.

[94] Lee, T. P., and C. Zah, "Wavelength-tunable and single-frequency semiconductor lasers for photonic communications networks," *IEEE Communications Mag.*, Vol. 27, No. 10, 1989, pp. 42–52.

[95] Henry, C. H. "Theory of linewidth in semiconductor lasers," *IEEE J. Quantum Electron.*, Vol. QE-18, No. 2, 1982, pp. 259–264.

[96] Saito, S. and Y. Yamamoto, "Direct observation of Lorentzian line shape of semiconductor laser and linewidth reduction with external grating feedback," *Electronics Letters*, Vol. 17, 1981, pp. 325–327.

[97] Petermann, K., *Laser Modulation and Noise*, Kluwer Academic, Boston, 1991.

[98] Lau, K. Y. and A. Yariv, "Intermodulation distortion in a directly modulated semiconductor injection laser," *Appl. Phys. Lett.*, Vol. 45, No. 10, 1984, pp. 1034–1036.

[99] Lee, T. P. "Recent developments in LED's for optical fiber communication systems," *Proc. Society of Photo-Optical Instrument Engineers*, Paper 224-16.

[100] Gimlett, J. L., M. Stern, R. S. Vodhanel, N. K. Cheung, G. K. Chang, H. P. Leblanc, and P. W. Shumate, "Transmission experiments at 560 Mb/s and 140 Mb/s using single mode fiber and 1300 nm LEDs, *Electronics Letters*, Vol. 21, p. 1198.

[101] Liu Y. S. and D. A. Smith, "The frequency response of an amplitude modulated GaAs luminescence diode," *Proc. IEEE*, Vol. 63, p. 542.

[102] Kaminow, I. P., G. Eisenstein, L. W. Stulz, and A. G. Dentai, "Lateral confinement InGaAsP superluminescent diode at 1.3 μm," *IEEE J. Quantum Electron.*, Vol. 19, 1983, p. 78–82.

[103] Pearsall, T. P., "Photodetectors for optical communication," *J. Opt. Commun.*, Vol. 2, 1981, pp. 42–48.

[104] Brian, M., and Lee, T. P., "Optical receivers for lightwave communication systems," *IEEE Journal of Lightwave Technology*, Vol. LT-3, no. 6, 1985, pp. 1281–1300.

[105] Forrest, S. R., "Optical detectors for lightwave communication," in *Optical Fiber Telecommunications II*, Chap. 14, Academic press, New York, 1988.

[106] Campbell, J. C., "Photodetectors for long-wavelength lightwave systems," in *Optoelectronic Technology and Lightwave Communication Systems*," Chap. 14, Van Nostrand Reinhold, New York, 1989.

[107] Melchior, H., M. B. Fisher, and F. R. Adams, "Photodetectors for optical communication systems," *Proc. IEEE*, Vol. 58, No. 10, 1970, pp. 1466–1486.

[108] Smith, R. G., "Photodetectors and receiver-an update," *Topics in Applied Physics: semiconductor devices for optical communication*, edited by Kressel, H., Vol. 39, Springer-Verlag, 1980, pp. 293–299.

[109] Personick, S. D., "Statistics of a general class of avalanche detectors with applications to optical communication," *Bell Sys. Tech. Journal*, Vol. 50, No. 10, 1972, pp. 3075–3095.

[110] McIntyre, R. J., "Multiplication noise in uniform avalanche diodes," *IEEE Trans. Electron Devices*, Vol. ED-13, No. 1, 1966, pp. 164–168.

[111] Kim, O. K., S. R. Forrest, W. A. Bonnor, and R. G. Smith, "A high gain InGaAs/InP avalanche photodiode with no tunneling leakage current," *Appl. Phys., Lett.*, Vol. 39, 1981, pp. 402–404.

[112] Capasso, F., R. A. Logan, A. Hutchinson, and D. D. Manchon, "InGaAsP/InGaAs heterojunction p-i-n detectors with low dark current and small capacitance for 1.3–1.6 μm fiber optic systems," *Electronics Letters*, Vol. 16, no. 23, 1980, pp. 893–895.

[113] Capasso, F., W. T. Tsang, A. L. Hutchinson, and G. F. Williams, "Enhancement of electron impact ionization in a superlattice: a new avalanche photodiode with large ionization rates ratio," *Appl. Phys. Lett.*, Vol. 40, 1980, pp. 38–40.

[114] Miller, C. M., S. C. Mettler, and I. A. White, *Optical Fiber Splices and Connectors*, Marcel Dekker, New York, 1986.

[115] Tsuneo, N. and Uchida, N. "Optical cable design and characterization in Japan," *Proc. IEEE*, Vol. 68, 1980, pp. 1220–1225.

[116] Dagleish, J., "Splices, connectors, and power couplers for field and office use," *Proc. IEEE*, Vol. 68, 1980, 1226–1231.

[117] Goodfellow, R. G., A. C. Carter, I. Griffth, and R. R. Bradley, "GaInAsP/InP fast high radiance 1.05–1.3 μm wavelength LEDs with efficient lens coupling to small numerical aperture silica optical fibers," *IEEE Trans. on Electron. Devices*, Vol. Ed-26, 1979, p. 1215.

[118] Wada, O. S., Yamakoshi, M. Abe, Y. Yishitoni, and T. Sakwai, "High radiance InGaAsP/InP lens LED's for optical communication systems," *IEEE J. Quantum Electron.*, Vol. QE-17, 1981, p. 174.

[119] Khoe, G. D., et al., "Progress in monomode optical-fiber interconnection devices," *IEEE Journal of Lightwave Technology*, Vol. LT-2, 1984, pp. 217–227.

[120] Suzuki, T., et al., "High speed 1.3 μm LED transmitter using GaAs driver IC," *IEEE Journal of Lightwave Technology*, Vol. LT-4, 1986, p. 790.

[121] Nakano, J., et al., "Monolithic integration of laser diodes, photomonitors, and laser driver circuits on a semi-conducting GaAs," *IEEE Journal of Lightwave Technology*, Vol. LT-4, 1986, p. 576.

[122] Pal, B. P., *Fundamentals of Fiber Optics in Telecommunication and Sensor Systems*, New York: John Wiley and Sons, 1992.

C*hapter 3*

Optical Amplifiers

As was shown in previous chapters, the transmission distance is an important requirement for high capacity long-haul communication systems. The transmission distance can be increased by using optoelectronic regenerators or optical amplifiers. The use of a optoelectronic regenerator requires optoelectronic conversion that is quite complex and incurs higher cost. In comparison with a regenerator, an optical amplifier can boost the power of lightwave signals without the need for optoelectronic conversion and yields low cost. More importantly, an optical amplifier is transparent to the data rate and modulation format, and it can accommodate many wavelength division multiplexed channels.

An optical amplifier can be broadly classified into two types: (1) the semiconductor laser amplifier (SLA) or (2) the doped-fiber amplifier. A SLA is based on the structure of a semiconductor laser, whereas, a doped fiber amplifier consists of a short length of optical fiber doped with rare earth ions such as erbium (Er^{3+}) or neodymium (Nd^{3+}) or praseodymium (Pr^{3+}) pumped with laser sources. An optical amplifier can be utilized as a pre-amplifier for the receiver or as a booster for the laser transmitter or as an in-line amplifier in a long-haul system.

In recent years, an erbium doped fiber amplifier has proved to be very attractive for many areas of optical fiber telecommunications: undersea systems, terrestrial long-haul networks, and telephone trunk lines. In comparison with a semiconductor laser amplifier, an optical fiber amplifier offers high gain, low splice loss and reflection, and low intermodulation distortion and is polarization insensitive. In addition, SLAs are not utilized in WDM systems because of the amplifier-induced crosstalk effects.

This chapter is devoted to optical amplifiers. In the following sections, we review their basic concepts, characteristics, and applications to optical fiber transmission systems [1–12]. Semiconductor laser amplifiers are discussed in Section 3.1, and Section 3.2 describes optical fiber amplifiers. The noise characteristics of these amplifiers are discussed in Section 3.3. The amplifier induced-crosstalk is discussed in Section 3.4, and finally, the system applications of optical amplifiers are described in Section 3.5.

3.1 SEMICONDUCTOR AMPLIFIERS

Semiconductor laser amplifiers are classified into two types: (1) the Fabry-Perot amplifier (FPA) and (2) the traveling wave amplifier (TWA). The schematic diagram of a typical semiconductor amplifier is shown in Figure 3.1(a). It is a double-heterostructure laser diode, with active region width W, thickness d, and length L. The mirror facets of the laser cavity are coated with antireflection coating. The incident light enters the FP cavity from the left side of the facet, passes through the active layer with a gain, and travels out from the right side of the facet. A portion of the light reflects from the right side of the facet and passes through the active layer and then travels out, leading to multiple reflections, whose resonance characteristics are shown in Figure 3.1(b).

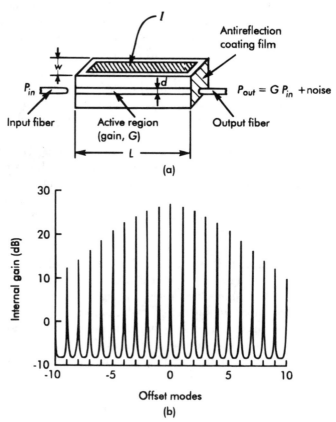

Figure 3.1 (a) Schematic representation of a typical semiconductor laser amplifier and (b) Fabry-Perot Resonance. Here, G is the amplifier gain, P_{in} is the input power, and P_{out} is the output power. (*Source:* [1]. © 1988 IEEE. Reprinted with permission.)

On the other hand, only a single path exists in the case of a TWA, which has very low mirror facet reflectivities. The factor $F = G_S\sqrt{R_1 R_2}$ is close to unity in the case of an FPA, and it is close to zero in the case of a TWA. In practice F ranges between 0.5 and 0.99 for FPA operation and the value of F for TWA operation is less than 0.05. However, near-TWA operation can be obtained with F in the range of 0.05–0.5. G_S denotes the single pass gain through the device, which is given by

$$G_S = \exp^{g(z)L} \tag{3.1}$$

where $g(z)$ denotes the gain of an amplifier in the active region during the single path. It is important to note that the gain depends on z, which represents the length of the active region.

The most promising semiconductor amplifiers are the TWAs based on the multiple quantum well structures, because these amplifiers have a flat gain spectrum, reduced polarization sensitivity, wide bandwidth, and high saturation output powers as compared to FPAs. However, TWAs require very low facet reflectivities to eliminate the Fabry-Perot cavity resonances. The feedback required for laser oscillations can be reduced by depositing dielectric anti-reflection coating film on the mirror facets. In recent years, good antireflection coating of less than 0.0001 have been obtained in InGaAsP semiconductor lasers.

The operation of semiconductor laser amplifiers is similar to that of semiconductor lasers, except the fact that laser amplifiers operate below threshold and use either low injection current or low facet reflectivities or both. The carriers are injected into the active region from an external bias source. If a sufficient number of carriers are injected, then an optical signal impinging on the active region will induce stimulated emission, and consequently the optical signal is amplified. The active region serves as the core of the waveguide in which the optical signal propagates along the device as it is amplified.

The resulting gain in the active region can be expressed in terms of the gain per unit length g as [1]

$$g(z) = g_m \Gamma_c - \alpha \tag{3.2}$$

where Γ_c is the optical confinement factor and α represents the effective loss coefficient per unit length and g_m denotes the material gain coefficient which can be approximated as

$$g_m = a_0(n(t) - n_0) \tag{3.3}$$

where a_0 is the gain constant and n_0 is the transparent carrier density. The carrier density and the optical density are related by the rate equation

$$\frac{dn(t)}{dt} = \frac{J}{qd} - \frac{n(t)}{\tau} - \frac{cg_m\Gamma_c}{n_r} S(z) \tag{3.4}$$

where J is the injected current density, q is the electronic charge, d is the thickness of the active region, Γ_c is the optical confinement factor, c is the velocity of light, τ is the carrier lifetime, $n(t)$ is the carrier density, and n_r denotes the material refractive index. The term $S(z)$ represents the optical density of the signal along the active region of the amplifier. Hence, the steady state solution of (3.4) is given by

$$\frac{J}{qd} = \frac{n(t)}{\tau} + \frac{cg_m\Gamma_c}{n_r} S(z) \tag{3.5}$$

For the case of zero optical signal, the resulting carrier density from (3.5) is $n(t) = n_1 = J\tau/qd$ as well as the corresponding material gain coefficient from (3.3) is

$$g_0 = g_m|_{n(t)=n_1} = a_0\left(\frac{J\tau}{qd} - n_0\right) \tag{3.6}$$

when the optical signal is greater than zero, different values of carrier density and gain will result. Hence, using (3.5), the material gain coefficient can be written as

$$g_m = \left(\frac{J}{qd} - \frac{n(t)}{\tau}\right)\frac{n_r}{cS(z)\Gamma_c} \tag{3.7}$$

Using (3.6), (3.7), and $n(t) = g_m/a_0 + n_0$ from (3.3), it can be shown that the material gain coefficient can be expressed as

$$g_m = \frac{g_0}{1 + P(z)/P_{sat}} \tag{3.8}$$

where $P(z) = h\nu cS(z)/n$ denotes the light intensity and $P_{sat} = h\nu/a\tau\Gamma_c$ represents the saturation intensity of the signal. From (3.8), it can be seen that the single pass gain is reduced by a factor of two when the light intensity $P(z)$ is equal to the saturation intensity P_{sat}.

The expression for the single pass gain $G_S = G$ with zero facet reflectivities can be obtained for an ideal TWA by assuming $\alpha = 0, \Gamma_c = 1$ in (3.2). Then from (3.8), we have

$$\frac{dP(z)}{dz} = \frac{P(z)g_0}{1 + P(z)/P_{sat}} \tag{3.9}$$

Integrating (3.9) over the amplifier length from $z = 0$ to L, it can be shown that the gain of an amplifier can be written as

$$G = 1 + \frac{P_{sat}}{P_{in}} \ln\left(\frac{G_0}{G}\right) \tag{3.10}$$

where $P_{in} = P(0)$ and $G = P(L)/P(0) = P_{out}/P_{in}$. The term $G_0 = e^{g_0 L}$ denotes the single pass gain of the amplifier in the absence of the signal. The dependence on the gain on the input power of an amplifier is plotted in Figure 3.2. From Figure 3.2, it can be seen that the gain of an amplifier decreases with an increase in signal input power, resulting in gain saturation.

The nominal phase shift associated with the single pass gain of an amplifier is given by [4]

$$\Phi = \frac{\pi(\nu - \nu_0)}{\Delta\nu} \tag{3.11}$$

where ν is the signal frequency, ν_0 denotes the resonant frequency of the FP cavity, and $\Delta\nu$ represents the longitudinal mode spacing. It is important to note that the gain and phase of an amplifier vary with the signal frequency and may result in inherent signal distortion. The resulting 3 dB bandwidth for the unsaturated amplifier is given by [1]

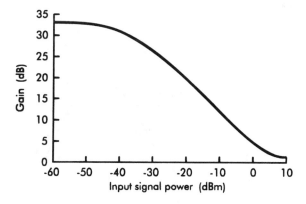

Figure 3.2 Single pass gain G_S response of a semiconductor TWA as a function of input power P_{in}, assuming saturation power $P_{sat} = -6$ dBm and a small signal gain $G_0 = 32.5$ dB. (*Source:* [8]. © 1990 IEEE. Reprinted with permission.)

$$\Delta\lambda = \sqrt{\frac{\ln 2}{a_2 \Gamma_c L}} \tag{3.12}$$

where a_2 is a constant. From (3.12), it can be seen that the bandwidth of an amplifier is independent of the absolute gain, but dependent on the device length.

Another important parameter in describing the performance of an amplifier is its saturation power. The saturation output power of an amplifier is the power at which the single pass gain has decreased by 3 dB, which is given by [1]

$$P_{OS} = \frac{Wdh\nu}{a_0 \tau \Gamma_c} \ln 2 \tag{3.13}$$

The saturation output power can be increased by decreasing the carrier life time or the optical confinement factor. Typically, the saturation output power of semiconductor laser amplifiers is in the range of 5–10 dBm.

In practice, the AR coated facets can exhibit some residual reflectivity that results in an FPA. Hence, the gain, bandwidth, and the saturation characteristics will differ from those of an ideal TWA. Figure 3.3 compares the typical gain characteristics of TWAs with FPAs. From Figure 3.3, it can be seen that the saturation output power (at which the signal gain is decreased by 3 dB from the unsaturated value) of a TWA increases with the signal gain, in contrast to the case of the FPA. For example, the saturation output power of a FPA at a 20 dB signal gain is −13 dBm, and for TWA, it is 20 dB larger than that of the FPA. With an FP cavity, the gain response of an amplifier is given by [1]

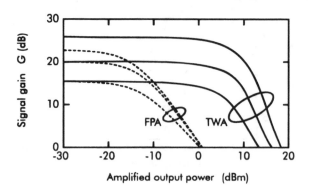

Figure 3.3 Gain characteristics of TWAs and FPAs. The solid and dashed curves denote the theoretical results for TWAs and FPAs respectively. (*Source:* [4]. © 1987 IEEE. Reprinted with permission.)

$$G = \frac{(1 - R_1)(1 - R_2)G_S}{(1 + \sqrt{R_1 R_2} G_S)^2 + 4\sqrt{R_1 R_2} G_S \sin^2 \Phi} \tag{3.14}$$

Note that the gain of an amplifier depends on the phase shift Φ, which is a function of signal frequency, FP cavity resonance frequency, and the longitudinal mode spacing. For ideal TWAs, the mirror facet reflectivities $R_1 = R_2 = 0$, then (3.14) reduces to G_S. The output saturation power of an FPA is less than that of TWA which is given by [1]

$$P_{OS} = \frac{Wdh\nu}{a_0 \tau \Gamma_c} \ln\left(\frac{2}{1 + RG_S}\right) \frac{(1 - R)G_S}{(1 + RG_S)(G_S - 1)} \tag{3.15}$$

where $R_1 = R_2 = R$ and $G_S \gg 1$. The corresponding 3 dB bandwidth can be expressed as a function of the cavity gain and mirror facet reflectvities as

$$\Delta\lambda = \frac{c}{2\pi n_r L} \sin^{-1}\left\{\frac{1}{2}\left[\frac{(1 - R_1)(1 - R_2)}{\sqrt{R_1 R_2} G}\right]^{1/2}\right\} \tag{3.16}$$

For example, with 25 dB gain and a cavity length of 200 μm, the bandwidth is on the order of 6 GHz. The amplifier gain also depends on the temperature and the state of polarization. For a near-TWA, the gain decreases by approximately 3 dB if the temperature is increased by 5°C, and conversely a decrease in the temperature increases the gain of an amplifier. In the case of polarization dependence, the single pass gain is different for TE and TM polarization modes, because of the different confinement factors. At a 25-dB gain, the difference is approximately 2.5 dB for the near-TWAs and 10 dB for the FPAs.

3.2 DOPED-FIBER AMPLIFIERS

The advent of doped-fiber amplifiers during the late 1980s revolutionized the field of lightwave communication systems and paved the way for long-distance communications. The typical configuration of a doped-fiber amplifier is shown in Figure 3.4. A doped-fiber amplifier consists of a short length of a single mode fiber (silica or fluoride) doped with rare earth ions (erbium or praseodymium or neodymium), which is pumped with a source at a lower wavelength (0.8, 0.98, and 1.48 μm) to provide population inversion. The optical signal which is propagating along the fiber is amplified from the stimulated emission of the excited ions. The erbium ions provide amplification in the 1.5 μm wavelength region and the praseodymium, and the neodymium ions provide amplification in the 1.3 μm wavelength region.

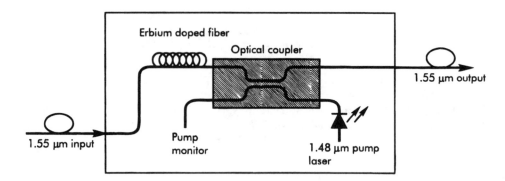

Figure 3.4 Configuration of a typical doped-fiber amplifier.

The light from the pump sources is coupled into the fiber from its front end (co-directional) or its near end (bi-directional) by using a WDM coupler. The isolators are used to suppress the laser oscillations and to prevent the feedback of amplified spontaneous emission noise (ASE). A filter can also be utilized at the output of an amplifier to obtain a flat gain response or to minimize the amplified spontaneous emission noise.

During recent years, the erbium doped-fiber amplifiers have proven to be very attractive due to their compact size, flattened gain spectrum, and reliable pump sources, and accordingly they are utilized in commercial 1.55 μm wavelength systems [9–13]. The praseodymium and neodymium doped-fiber amplifiers have been studied recently for the minimum dispersion 1.3 μm wavelength systems [14–19]. Other fiber amplifiers such as Fiber Raman and Brillouin amplifiers can also be realized by exploiting the fiber nonlinearity. In the following sections, we focus on the operating principles and characteristics of doped fiber amplifiers for 1.55 and 1.3 μm systems.

3.2.1 1.55 μm Fiber Amplifier

During recent years, the erbium-doped fiber amplifier (EDFA) has become an obvious choice for 1.55 μm systems. The basic principle of operation of an EDFA can be understood with the energy level diagram of an erbium ion in a silica fiber as shown in Figure 3.5. In Figure 3.5, an erbium ion is approximated with a three-level energy system. An erbium concentration of about 100 ppm is desirable for the fabrication of efficient EDFAs. The energy level designated by $^4I_{15/2}$ is called the ground state, the level of $^4I_{13/2}$ is referred to as the metastable state and the level of $^4I_{11/2}$ is known as the final state. The typical wavelength transitions are also included. The signal absorption from the ground state is often referred to as ground-state absorption (GSA).

One of the basic requirements for the operation of a fiber amplifier is pumping ions to a higher level. The 0.98 μm laser source acts as a "pump" to increase the

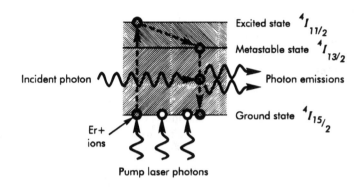

Figure 3.5 Three-level energy diagram of an erbium ion in silica fiber.

energy of an incoming signal as it travels along the fiber. The pump light excites the Er^{3+} ions to higher energy levels than their normal state, so when the signal light encounters the Er^{3+} ions, stimulated light emission causes signal amplification.

The process that an ion in the lower level state is excited to a higher level through the absorption of a pump photon, is often referred to as excited state absorption (ESA). In this process, some of the ions may return to the metastable state and be available for either spontaneous or stimulated emission. Excited ions have an ESA spectrum different from their GSA spectrum due to the new energy gaps between the ground states and the higher states. The transition time from the final state toward the metastable state is on the order of few μs.

An important factor in the operation of an EDFA is the long lifetime of the metastable state, which permits the required high-population inversions using the modest pump powers. The typical lifetime of the metastable state is on the order of 10 ms, and hence high amounts of energy remain stored in that level. In addition, an EDFA must have a suitable pump band so that the absorption bands coincide with the emission wavelengths of the pump sources.

The semiconductor lasers are used as pump sources in the 0.8-, 0.98-, and 1.48-μm absorption bands. Although the erbium ions can be pumped at several wavelengths to obtain population inversion for the 1.55-μm emission, only the pump bands at 0.98 and 1.48 μm are shown in the energy level diagram. For pumping at the 0.98-μm band, AlGaAs semiconductor lasers are utilized and for pumping at the 1.48-μm band, InGaAs quantum well lasers are used.

EDFAs can be excited around 0.8 μm using reliable and low-cost AlGaAs laser diodes. However, pumping at 0.8 μm requires about 7–8 dB higher power than the other bands due to the excited state absorption, resulting in low efficiency. The pumping bands at 0.98 and 1.48 μm have been demonstrated to be very efficient for the EDFA, but the cost of the laser diodes is quite high in comparison with 0.8-μm laser diodes. The most widely used pump wavelength today is at 1.48 μm, because this band is free from ESA and the pump laser diode technology is more mature and reliable than a 0.98-μm pump band.

The cost and reliability of the pump sources are a strong function of the power level at which they are required to operate. Therefore, it is important to compare the power requirements of these three bands to achieve the same level of amplifier performance for different applications. The minimum required pump power for small-signal amplifiers and for booster applications is shown in Figure 3.6. Curves are shown for bi- and codirectional pumping in the 0.8-μm band and codirectional pumping at 0.98 and 1.48 μm. As seen from Figure 3.6, bidirectional pumping requires about 1 dB less power than the codirectional pumping at 0.8 μm. Also, it can seen that the required pump power for 0.98 μm is 2 dB lower than that for 1.48-μm pumping and 7 to 8 dB lower than that needed in the 0.8-μm pump band to provide the same small signal gain. Hence, the most efficient pump band is 0.98 μm for small-signal amplifier applications.

To compare the three pump bands for booster applications, where the objective is to maximize the signal output power, the curves are depicted in Figure 3.6(b). In this figure, the minimum required pump power is plotted as a function of the amplifier signal power output, assuming a signal input power of 1 mW. From Figure 3.6(b), it can be seen that for a signal power output of above 5 dBm, the 1.48-μm pump band requires the lowest pump power as compared to 0.8- and 0.98-μm bands because of the difference in energy per photon. For booster applications, 1.48 μm is more efficient than 0.8- and 0.98-μm bands.

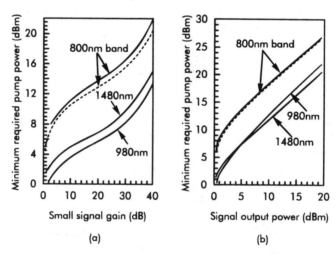

Figure 3.6 (a) Minimum required pump power versus small signal gain for different pump bands and (b) minimum required pump power versus signal output power for different pump bands. In both figures a step-index fiber with an NA = 0.25 and uniform erbium doping was used. (*Source:* [20]. © 1992 IEEE. Reprinted with permission.)

For preamplifier applications, in addition to the gain, a low noise figure is very important. The noise figure of an amplifier is a function of the fiber length, pump power and the pumping scheme. For the 0.8-μm band, when the noise figure is below 4 dB, the best pump wavelength is 0.805-μm and the pumping scheme should be co-directional [21]. However, the 0.8-μm band requires higher pump power than the 0.98-μm band. The required pump power as a function of the amplifier noise figure is shown in Figure 3.7 for all three bands. From Figure 3.7, it can be seen that the resonant pumping at 1.48 μm leads to a minimum noise figure of 4.8 dB for a signal wavelength of 1.532 μm, and consequently noise figures of less than 5 dB are not practical. With regard to 0.8- and 0.98-μm bands, the minimum noise figure is 3 dB for Gain >> 1. From Figure 3.7, it is observed that a significant improvement can be obtained by a slight increase the pump power for the minimum required power. It can be seen that 0.98 μm is the best pump wavelength for preamplifier applications, since 7–8 dB less power is required to obtain the same amplifier performance.

The gain and saturation characteristics of an EDFA can be determined from the rate equations, which in general involve solving the coupled differential equations by using numerical methods. In recent years, several models have been presented in the literature [22,23] to predict the amplifier characteristics. Many of these models originate from the rate and propagation equations but differ with respect to complexity. An outline of these models is presented in [11].

We use a simplified model that can describe the gain and saturation characteristics of a medium gain EDFA. This model gives an analytical expression for the gain

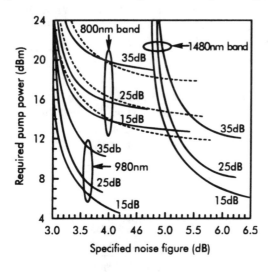

Figure 3.7 Required pump power as a function of specified noise figure for various small signal gains of 15, 25, and 30 dB respectively. (*Source:* [20]. © 1992 IEEE. Reprinted with permission.)

of an amplifier and an insight into the amplifier behavior. From the analysis of [24], the amplifier gain can be calculated by using

$$\frac{P_p^{in}}{h\nu_p}\left[1 - \left(\frac{G}{G_{max}}\right)^{\partial}\right] = \frac{P_S^{sat}}{h\nu_S}\left[\alpha_S L + \ln(G)\right] + \frac{P_S^{in}}{h\nu_S}(G - 1) \qquad (3.17)$$

The term on the left-hand side of (3.17) denotes the number of absorbed pump photons per unit time, and the terms on the right-hand side correspond to the rate of spontaneously emitted photons and the number of photons added to the signal per unit time, respectively. The terms P_p^{in} and P_s^{sat} are the input powers of the pump and signal respectively. The saturation power of the signal or the pump ($k = s$ for the signal or $k = p$ for the pump) is defined by

$$P_k^{sat} = I_k^{sat}\frac{\int N_e\, dA \int f_k\, dA}{\int N_e f_k\, dA} \qquad (3.18)$$

where I_k^{sat} represents the saturation intensity

$$I_k^{sat} = \frac{h\nu_k}{(\sigma_k^{em} + \sigma_k^{abs})\tau} \qquad (3.19)$$

where σ_k^{em} and σ_k^{abs} are the emission and absorption cross section of the spectrum respectively. The term N_e denotes the local erbium concentration. The term f_k denotes the power mode field function, which can be approximated by a Gaussian profile. The maximum gain G_{max} in (3.17) is given by

$$G_{max} = e^{(\alpha_p/\partial - \alpha_S)L} \qquad (3.20)$$

where the small-signal absorption coefficients α_P and α_S are represented by α_k ($k = s$ for the signal and $k = p$ for the pump), which can be estimated from

$$\alpha_k = \sigma_k^{abs}\frac{\int N_e f_k\, dA}{\int f_k\, dA} \qquad (3.21)$$

and the ratio of saturation powers is defined by

$$\partial = \frac{P_S^{sat}/h\nu_S}{P_P^{sat}/h\nu_p} \qquad (3.22)$$

and the term L in (3.20) denotes the fiber length. It is important to note that the gain equation (3.17) is applicable to medium gain amplifiers (about 20 dB or with high signal input powers of greater than −20 dBm) that are not saturated by the amplified spontaneous emission noise. The gain equation is plotted in Figure 3.8 as a function of the signal input power for the power at 0.98 μm. From Figure 3.8, it can be seen that the gain decreases with the increase in signal input power, as in the case of semiconductor laser amplifiers. The saturation characteristics of EDFAs are similar to those of the semiconductor laser amplifiers. Practical EDFAs with the saturation powers of around 500 mW and output powers of about 100 mW are available commercially. For high signal powers, the gain can be determined from [24]

$$G = 1 + \frac{\lambda_P P_P^{in}}{\lambda_S P_S^{in}} \tag{3.23}$$

In this case, the gain is approximately independent of the signal wavelength.

3.2.2 1.3-μm Doped-Fiber Amplifiers

During the last decade, major research was focused on an erbium doped silica fiber because of its potential as a high-gain optical amplifier for the 1.5-μm systems. However, the majority of installed optical communication systems operate at 1.3 μm. Therefore, the recent development of 1.3-μm amplifiers is of great interest. These amplifiers fall under a four-level system and the pumping band is at wavelengths anywhere between 0.8 and 1.05 μm. These amplifiers are still in development and

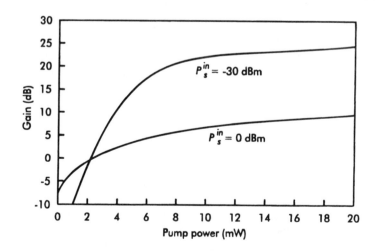

Figure 3.8 (a) Gain characteristic of an EDFA for 0.98 μm pumping for different signal input powers. (*Source:* [24]. © 1992 IEEE. Reprinted with permission.)

most of the current research is focused on their performance improvements and optimization.

As a first candidate, Nd^{3+}-doped silica fibers were tried for the amplification at 1.3 μm systems, but they have been shown to reveal an ESA near 1.3 μm and a strong ASE at 1.05 μm [14]. The ESA spectrum is sensitive to glass composition and in fluoride glasses, the spectrum is shifted downward in amplitude and wavelength and thus creates amplification. Hence, using the fluorzirconate glass ZBLANP (Zr, Ba, La, Al, Na, P) composition fibers, the gain of up to 7 dB at 1.34 μm has been achieved [15].

Since the gain maximum is located around 1.34 μm and the gain overlap with the 1.3-μm telecommunication window is relatively poor, a fundamental problem exists for the upgrading of the existing 1.3 μm systems. Gain improvements of a few more decibels can be achieved by filtering out the ASE emitted at 1.05 μm [16]. In summary, the performance characteristics of these neodymium-doped ZBLANP fiber amplifiers yield results far below the values obtained by erbium-doped amplifiers.

Another promising candidate for 1.3-μm systems is a praseodymium rare earth dopant. A praseodymium concentration in fluoride fiber of less than 1000 ppm is desirable for the fabrication of efficient praseodymium fluorozirconate fiber amplifiers (PDFAs). For high gain, the pumping power is in the range of 300–1000 mW, which is higher than the pump power required for EDFAs. The performance of these amplifiers is limited by the ESA band peaking close to 1.38 μm and the GSA band that peaks around 1.4 μm.

In recent years, the PDFAs have provided a gain values as high as 38.2 dB for a pump power of 300 mW, but with a low quantum efficiency [17,18]. The 3 dB bandwidth of an amplifier is in the range of 30–40 nm and the typical output saturation power is in the range of 10–12 dBm. In the near future, PDFAs may offer the best prospects for 1.3 μm operation, and the future might reveal new combinations of both the host glass and rare earth compositions for 1.3 μm amplifiers.

The basic characteristics of semiconductor as well as doped-fiber amplifiers are compared in Table 3.1.

3.3 NOISE CHARACTERISTICS

Besides the gain and saturation characteristics, another main parameter of an optical amplifier is the noise figure, which is denoted by NF. The amplification of an optical signal is associated with the addition of a broadband amplified spontaneous emission noise (ASE) to the signal. The ASE can be quantified by the noise figure of an optical amplifier, which in turn can be used to calculate the effect of ASE on the received signal to noise ratio (SNR). The noise figure of an amplifier is defined as

Table 3.1
Basic Characteristics of Optical Amplifiers

Characteristics	TWA	EDFA	PDFA
Wavelength	0.8, 1.3, 1.5 μm	1.55 μm	1.3 μm
Bandwidth	30–40 nm	30–40 nm	30–50 nm
Gain	20–25 dB	30–45 dB	20–30 dB
Saturation power	5–10 dBm	5–10 dBm	10–12 dBm
Polarization Sensitivity	Yes	No	No
WDM Crosstalk	Yes	No	No
Pump wavelength	N/A	0.8, 0.98, 1.48 μm	0.95–1.05 μm
Pump powers	N/A	20–100 mW	100–1000 mw

Note: The pump powers indicated in the Table 3.1 reflect some of the typical sources and are not correlated with the amplifier gain and saturation values.

$$NF = \frac{SNR_{\text{in}}}{SNR_{\text{out}}} \tag{3.24}$$

Where SNR_{in} denotes the SNR (electrical) at the front end of an amplifier and SNR_{out} represents the SNR (electrical) at the output of an amplifier. The significance of (3.24) can be understood from the fact that higher NF of an amplifier deteriorates the SNR_{out} of an amplifier. An accurate description of noise in semiconductor and doped fiber amplifiers is described in [6,7] and [25–28] respectively.

In general, the noise figure of an optical amplifier depends on the spectral density of the ASE, the amplifier gain, and the photodetector parameters. However, a simple expression for the NF can be obtained by assuming an ideal photodetector with unit quantum efficiency, whose performance is limited by the beat noise between the signal and ASE. Let us consider an amplifier with a gain G, whose power spectral density of the ASE noise is approximated as

$$S_{\text{ASE}}(f) = h\nu n_{\text{sp}}(G - 1) \qquad 0 < f < \infty \tag{3.25}$$

where n_{sp} is called the spontaneous emission parameter (also known as population inversion parameter) of an amplifier. It is important to note that in the case of doped fiber amplifiers, n_{sp} is strongly dependent on the fiber length and the pumping scheme. The SNR_{in} of an amplifier is given by the ratio of average signal power to the noise power incurred by the detector shot noise as

$$SNR_{\text{in}} = \frac{(\Re P_{\text{in}})^2}{2q\Re P_{\text{in}}B} = \frac{P_{\text{in}}}{2h\nu B} \tag{3.26}$$

where $\Re = q/h\nu$ denotes the responsivity of the photodetector, P_{in} is the input power, and B is the photodetector bandwidth. Similarly, the output SNR of an amplifier is given by the ratio of the output signal power to the output

$$SNR_{out} = \frac{(\Re GP_{in})^2}{4\Re^2 GP_{in}S_{ASE}B}.$$ (3.27)

where the term in the denominator represents the beat noise between the signal and ASE, which is assumed to be the dominant noise term. The thermal noise, shot noise, and the spontaneous-spontaneous noise components are assumed to be negligible. Equation (3.27) is simplified to

$$SNR_{out} \approx \frac{GP_{in}}{4S_{ASE}B}$$ (3.28)

Using (3.28), (3.26), and (3.25) in (3.24), the noise figure of an amplifier can be written as

$$NF = \frac{2n_{sp}(G - 1)}{G} \approx 2n_{sp}$$ (3.29)

The noise figure has a minimum value of 3 dB for a high gain fully inverted amplifier with $n_{sp} = 1$. In practical semiconductor laser amplifiers, the NF varies in the range of 6–8 dB [6,7]. For doped-fiber amplifiers, noise figure varies from 4–6 dB [25–28].

In the case of a cascaded amplifier system, the spontaneous emission noise of each optical amplifier propagates along the fiber and is amplified by the successive amplifiers. Thus ASE can accumulate before reaching the receiver, and as the ASE level increases, it can saturate the optical amplifiers and hence decrease the SNR in the system. Let us consider the cascaded system as shown in Figure 3.9, in which the amplifiers and interamplifier attenuations are represented as a single component

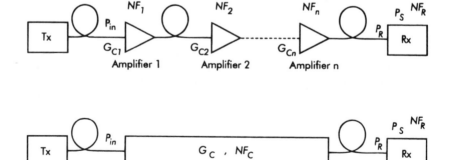

Figure 3.9 Amplifier cascade model. (*Source:* [29]. © 1991 IEEE. Reprinted with permission.)

with cumulative gain G_C and cumulative noise figure NF_C. The cumulative gain of the cascade is the product of the individual amplifier gains and interamplifier attenuations, whereas the cumulative noise figure NF_C in a cascaded amplifier system is given by [29]

$$NF_C = NF_1 + \frac{NF_2}{G_{C2}} + \cdots + \frac{NF_n}{G_{Cn}} \tag{3.30}$$

where NF_i ($i = 1 \cdots n$) denotes the noise figure of the ith amplifier and G_{Ci} represents the cumulative gain of the cascade up to the input of the ith amplifier. In terms of the cascade noise figure, the system requirement on minimum input power is given by [29]

$$\frac{NF_C}{P_{in}} + \frac{NF_R}{P_R} \leq \frac{1}{2}\frac{NF_R}{P_S} \tag{3.31}$$

where P_{in} represents the input power to the cascade, P_R is the received power, NF_R is the noise figure of the receiver, and P_S denotes the receiver sensitivity for a specified BER in the absence of an amplifier. The factor of a half is introduced as a practical SNR margin for 3 dB. For example, a transmission system operating with a cumulative noise figure of 15 dB along with P_S/NF_R of -46 dBm would require a minimum input power of -46 dBm. Hence, the input power to the cascade system is dependent on the receiver sensitivity, which in turn depends on the system bit rate. It is important to note that the cumulative gain and noise figure of the cascade system decrease with the increase in saturation power [29]. For a proper system operation, the required number of cascaded amplifiers have to be estimated

3.4 AMPLIFIER-INDUCED CROSSTALK

As mentioned in the introduction, an optical amplifier is an attractive device for multichannel systems, because of its ability to amplify a number of multiplexed channels simultaneously. However, the crosstalk between the channels (interchannel crosstalk) can be induced by an amplifier in a multichannel system due to the nonlinear gain characteristics of SLAs. Specifically, near saturation, an increase in the input signal power reduces the carrier density in the active region by stimulated emission and thus changes the refractive index of an amplifier. This reduces the gain of an amplifier, which affects the power of neighbor channels, and results in cross-saturation. In this case, the signal gain of one channel changes from bit to bit, and the change depends on the bit pattern of neighboring channels, which is undesirable for ASK or OOK lightwave systems. The crosstalk induced by cross-saturation can be minimized by operating SLAs in the unsaturated region of the amplifier.

Another factor of crosstalk in SLAs is the carrier density modulation (four-wave mixing) at beat frequencies of the incident signals. This crosstalk mechanism is strongly dependent on channel frequency separation and can be avoided for relatively large channel spacing of greater than 10 GHz. A detailed analysis of crosstalk effects in semiconductor laser amplifiers is presented in several articles [8,30–32]. In the case of doped fiber amplifiers, it has been found that the crosstalk is heavily dependent upon the modulating frequency and found to be negligible at frequencies above 100 kHz [33]. Therefore, for WDM systems, erbium-doped amplifiers are preferred.

3.5 SYSTEM APPLICATIONS

As already mentioned in the introduction, an optical amplifier can be utilized in several configurations in lightwave systems. These configurations are shown in Figure 3.10. In addition to these configurations, SLAs are used as switches, nonlinear, and

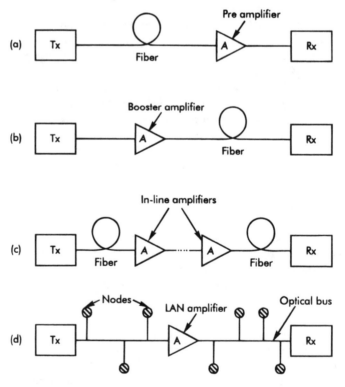

Figure 3.10 (a) Pre-amplifier; (b) booster amplifier; (c) in-line amplifier; and (d) LAN system. (*Source:* [42]. Reprinted with the permission of John Wiley and Sons.)

bistable elements for wavelength conversion and TDM demultiplexing. For example, the nonlinear behavior of an SLA can be used for pulse shaping, such that a dispersed optical pulse can be reshaped into a rectangular pulse in regenerator applications [1]. The use of an SLA as an optical switch will have an important role in future optical networks, for routing the optical signals by the application of electrical bias current [1]. In this section, we discuss the system design requirements for each configuration. We will, however at this point assume only EDFAs in these applications.

3.5.1 Preamplifier

Figure 3.10(a) illustrates the application of an optical amplifier as a preamplifier for the receiver in a lightwave system. In this case, the receiver sensitivity depends on the amplifier gain, noise figure, and optical bandwidth. Hence, an optical amplifier should have a low noise figure and a high gain so that the receiver thermal noise can be negligible, yielding higher receiver sensitivity. Furthermore, the signal power available at the input of an amplifier is usually low and therefore the saturation power of the amplifier need not be very large.

3.5.2 Booster Amplifier

Figure 3.10(b) depicts the use of an optical amplifier as a booster (also known as a power amplifier) for the transmitter. In this application, an amplifier should have high saturation power and not necessarily large gain.

3.5.3 In-line Amplifier

The use of an optical amplifier as an-in line amplifier is shown in Figure 3.10(c), where the amplifier compensates for the losses in the system. The losses may be due to the fiber losses in a long haul system, tap and splitting losses in a LAN, or coupling and switch losses in an optical switch. Let us consider the long-haul system. For this application, the main requirements are the amplifier placement and the number of in-line amplifiers in a cascaded system.

First we will address the question: what is the optimum placement of an amplifier in the fiber line to achieve the specified BER? To address this question first consider placing the amplifier near the transmitter, so that the ASE noise of the amplifier is sufficiently attenuated before reaching the receiver. The amplifier can also be inserted near the receiver to overcome the thermal noise of the receiver and to minimize the amplifier saturation. In this case, the ASE of the amplifier will be the dominant noise. In this situation, an optimum point along the link can be chosen to optimize the system performance parameters. The optimum placement of a single amplifier is in the range of $L/L_{tot} = 0.4$ to 0.6, where L_{tot} represents the total length of the lightwave link [34]. Additionally, the in-line amplifier transmission distance is affected by the fiber nonlinear effects such as self-phase modulation and group velocity dispersion [35].

Next let us address the second requirement: what is the maximum number of in-line amplifiers in a cascaded link? In this case, several identical amplifiers are placed equidistantly with a spacing of L, yielding a total link length of $L_T = N_A L$ where N_A denotes the chain of amplifiers. In a cascaded link, the ASE noise from each amplifier accumulates and hence can limit the performance of the receiver. A rough estimate of the maximum number of amplifiers is determined by choosing the spacing L such that the gain of each amplifier compensates the link loss between them. Hence, we have

$$Ge^{-\alpha L} = 1 \tag{3.32}$$

where α is the loss coefficient of the fiber. Then using $L = L_T/N_A$ in (3.32) and upon simplification, we get,

$$N_A = \frac{L_T \alpha}{\ln G} \tag{3.33}$$

The field trials and experiments of the in-line amplifier systems in direct-detection, coherent detection, and soliton systems are discussed in Chapters 4, 5, and 6, respectively.

3.5.4 WDM Systems

An optical amplifier is an ideal device in multichannel lightwave systems, as bulk amplification of high density WDM signals is possible with minimum crosstalk and yields higher system capacity. For example, the gain provided by a single EDFA stage increases the fiber link length from 100 to 150 km. In the case of a cascaded system, the channel capacity of the link is limited by the noise and saturation performance of the amplifier cascade. Equation (3.31) sets a minimum power per channel at the input of an amplifier cascade and at the same time the maximum total input power is also restricted.

In practice, the maximum total input power is limited by considerations such as interchannel crosstalk, finite transmitter power, saturation power of the individual amplifiers, and fiber nonlinear effects. Hence, the maximum operating channel capacity of a cascade WDM system can be found from (3.31) as

$$N \leq \frac{1}{2} \frac{P_{\max}}{NF_C} \left(\frac{P_S}{NF_R} \right)^{-1} \tag{3.34}$$

where N is the number of identical WDM channels and allowance is again made for a 3 dB operating margin at the input to the cascade. The channel capacity of a cascaded system is directly proportional to ratio of the maximum input power to the cumulative noise figure. Additionally, uniform gain spectrum, gain saturation, amplifier bandwidth, and a low noise figure of each amplifier are important for WDM applications.

In recent years, several optical channels have been multiplexed and distributed using EDFAs [36,37]. The system design issues include flattening of the EDFA gain spectrum, gain spectrum imbalance due to gain competition in gain saturated EDFAs, noise figure dependence on the pump and signal powers, and possible methods for widening the gain spectrum. Among these issues, EDFA nonuniform gain is the most difficult issue to solve. In this case, the EDFA gain spectrum is wavelength-dependent whereas the link loss between amplifiers is wavelength-independent (to the first order). This gain/loss wavelength indiscrepancy will result in a severe SNR differential among many channels after passing through a cascade of EDFAs and may cause system impairment.

In recent years, several methods have been used to correct the nonuniformity of the EDFA gain. Some of the methods utilized attenuators, smoothing filters such as Fabry-Perot or tunable Mach-Zehnder interferometers [38,39]. Another method was based on a system telemetry, where the output SNR of all the WDM channels can be equalized by adjusting the individual input signal powers with variable attenuators [40], which yields a good uniformity than the other methods. Additionally, an equalized SNR can be achieved by incorporating a notch filter (having a 3 dB bandwidth of 2 nm and a center wavelength of 1.56 µm after every group of 20 EDFA's without a priori knowledge of the input or output signals [41]. The use of optical amplifiers in WDM systems along with the field trials and experiments is discussed in Chapter 8.

Problems

Problem 3.1

Derive (3.8) and (3.10).

Problem 3.2

A 300 µm long semiconductor laser is used as an FP amplifier. Calculate the gain of an amplifier by assuming 32% reflectivity for both facets and a 30-dB single-pass gain. The nominal phase shift of an amplifier is 10 degrees. Also, find the 3 dB bandwidth of an amplifier. Assume the refractive index of the active region as 1.4.

Problem 3.3

Calculate the gain of a traveling amplifier if $G_0 = 25$ dB, $P_{in} = -20$ dBm, and $P_{sat} = 5$ dBm. Also, find the 3 dB bandwidth of an amplifier given $a_2 = 0.15/cm.nm^2$, $\Gamma_c = 0.3$, and $L = 500$ μm.

Problem 3.4

A traveling wave amplifier can amplify a 1 μW signal to the 1 mW level. What is the output power, when a 1 mW signal is incident on the same amplifier? The saturation power of an amplifier is 10 mW.

Problem 3.5

Calculate the gain of an EDFA given the following data. Assume a step index single mode fiber with the core radius $a = 1.4$ μm and $NA = 0.26$. Further data are: $\lambda_P = 0.98$ μm, $\lambda_S = 1.55$ μm, $\sigma^{abs} = \sigma^{abs} = 2 \times 10^{-21}$ cm^2, $\sigma_P^{em} = 0$, $\sigma_S^{em} = 3 \times 10^{-21}$ cm^2, and $\tau = 10$ ms. The erbium concentration is $N_e = 1.5 \times 10^{18}$ cm^{-3}. Assume the signal input power as -30 dBm.

Problem 3.6

Calculate the noise figure of a 1.55-μm amplifier with a gain of 35 dB. Assume that the spontaneous emission parameter of the amplifier is 1.4.

Problem 3.7

Calculate the overall noise figure of a three stage amplifier system, having the following parameters $NF_1 = 3$ dB, $NF_2 = 4$ dB, $NF_2 = 6$ dB, $G_{C2} = G_{C3} = 25$ dB.

Problem 3.8

For the system given in Problem 3.7, find the minimum input power required given $P_S/NF_R = -40$ dBm and $P_R/NF_R = -30$ dBm.

Problem 3.9

Calculate the number of in-line amplifiers for a system having a span length of 100 km, $G = 25$ dB, and $\alpha = 0.2$ dB/km.

Problem 3.10

Calculate the maximum total input power for a cascaded WDM system having 100 channels if $NF_C = 15$ dB and $P_S/NF_R = -40$ dBm.

References

[1] O'Mahony, M. J., "Semiconductor laser optical amplifiers for use in future fiber systems," *IEEE Journal of Lightwave Technology.*, Vol. LT-6, No. 4, 1988, pp. 531–544.

[2] Eisenstein, G., "Semiconductor optical amplifiers," *IEEE Circuits and Devices Magazine.*, Vol. 5, No. 4, pp. 25–30, 1989.

[3] Simon, J. C., "GaInAsP semiconductor laser amplifiers for single mode fiber communications," *IEEE Journal of Lightwave Technology.*, Vol. LT-5, No. 9, 1987, pp. 1286–1295.

[4] Tadashi, S. and T. Mukai, "1.5 μm GaInAsP traveling-wave semiconductor laser amplifier," *IEEE Journal of Quantum Electronics*, VOL. QE-23, No. 6, 1987, pp. 1010–1020.

[5] Yamamoto, Y., "Characteristics of AlGaAs Fabry-Perot cavity type laser amplifiers," *IEEE Journal of Quantum Electronics*, Vol. QE-16, 1980, pp. 1047–1052.

[6] Simon, J. C. "Semiconductor laser amplifier for single mode optical fiber communications," *Journal of Optical Communications*, Vol. 4, No. 2, 1983, pp. 51–62.

[7] Yamamoto, Y., "Noise and error rate performance of semiconductor laser amplifiers in PCM-IM optical transmission systems," *IEEE Journal of Quantum Electronics.*, Vol. QE-16, No. 10, 1980, pp. 1073–1081.

[8] Ramaswami, R. and P. A. Humblet, "Amplifier induced crosstalk in multichannel optical networks," *IEEE Journal of Lightwave Technology*, Vol. 8, no. 12, 1990, pp. 1882–1896.

[9] Special issue on optical amplifiers, *IEEE Journal of Lightwave Technology*, Vol. LT-9, No. 2, 1991.

[10] Special issue on system and network applications of optical amplifiers, *IEEE Journal of Lightwave Technology*, Vol. LT-13, No. 5, 1995.

[11] Bjarklev, A., *Optical fiber amplifiers*, Aretech House, Boston, 1993.

[12] Desurvire, E., *Erbium-doped fiber amplifiers*, John Wiley & Sons Inc., New York, 1993.

[13] Miniscalo, W. J., "Erbium-doped glasses for fiber amplifiers," *IEEE Journal of Lightwave Technology*, Vol. 9, no. 2, 1991, pp. 234–250.

[14] Dakss, M. L. and W. J. Miniscalco, "Fundamental limits on Nd-doped fiber amplifier performance at 1.3 μm," *IEEE Photonics Technology Letters*, Vol. 2, No. 9, 1990, pp. 650–652.

[15] Pedersen, J. E., M. C. Brierley, S. F. Carter, and P. W. France, "Amplification in the 1300 nm telecommunications window in a Nd-doped fluoride fiber," *Electronics Letters*, Vol. 26, No. 24, 1990, pp. 2042–2044.

[16] Obro, M., B. Pedersen, A. Bjarklev, J. H. Povlsen, and J. E. Pedersen, "Highly improved fiber amplifier for operation around 1300 nm," *Electronics Letters.*, Vol. 27, No. 5, 1991, pp. 470–472.

[17] Whitley, T. J., "A review of recent system demonstrations incorporating 1.3 μm praseodymium-doped fiber amplifiers," *IEEE Journal of Lightwave Technology*, Vol. 13, No. 5, 1995, pp. 744–760.

[18] Ohishi, Y., T. Kanamori, T. Nishi, and S. Takahashi, "A high gain, high output saturation power Pr-doped fluoride fiber amplifier operating at 1.3 μm," *IEEE Photonics Technology Letters*, Vol. 3, No. 8, 1991, pp. 715–717.

[19] Pedersen, B., W. J. Miniscalco, and R. S. Quimby, "Optimization of Pr-ZBLAN fiber amplifiers," *IEEE Photonics Technology Letters*, Vol. 4, No. 5, 1992, pp. 446–448.

[20] Pedersen, B., et al., "Power requirements for EDFA in the 800, 900, and 1480 nm bands," *IEEE Photonics Technology Letters*, Vol. 4, No. 1, 1992, pp. 46–48.

[21] Pedersen, B., et al., "Evaluation of the 800-nm pump band for erbium-doped fiber amplifiers," *IEEE Journal of Lightwave Technology*, Vol. 10, No. 8, 1992, pp. 1041–1049.

[22] Giles, C. R., and E. Desurvire, "Modeling of erbium-doped fiber amplifiers," *IEEE Journal of Lightwave Technology*, Vol. 9, No. 2, 1991, pp. 1443–1445.

[23] Pedersen, B., A. Bjarklev, J. H. Povlsen, K. Dybdal, and C. C. Larsen, "The design of erbium-doped fiber amplifiers," *IEEE Journal of Lightwave Technology*, Vol. 9, No. 9, 1991, pp. 1105–1112.

[24] Pfeiffer, T. and H. Bulow, "Analytical gain equation for erbium-doped fiber amplifiers including mode field profiles and dopant distribution," *IEEE Photonics Technology Letters*, Vol. 4, No. 5, 1992, pp. 449–451.

[25] Yamada, M., M. Shimizu, M. Okayasu, T. Takeshita, M. Horiguchi, Y. Tachikawa, and E. Sugita, "Noise characteristics of Er-doped fiber amplifiers pumped by 0.98 and 1.48 μm laser diodes," *IEEE Photonics Technology Letters*, Vol. 2, No. 3, 1990, pp. 205–207.

[26] Desurvire, E., "Spectral noise figure of Er-doped fiber amplifiers," *IEEE Photonics Technology Letters*, Vol. 2, No. 3, pp. 208–210, 1990.

[27] Pedersen, J. E., M. Brierley, and R. A. Lobbett, "Noise characterization of a Nd-doped fluoride fiber amplifier and its performance in a 2.4 Gb/s system," *IEEE Photonics Technology Letters*, Vol. 2, No. 10, 1990, pp. 750–752.

[28] Sdugawa, T., and Y. Miyajima, "Noise characteristics of Pr-doped fluoride fiber amplifier," *Electronics Letters*, Vol. 28, No. 3, 1992, pp. 246–247.

[29] Walker, G. R., N. G. Iker, R. C. Steele, M. J. Creaner, and M. C. Brain, "Erbium-doped fiber amplifier cascade for multichannel coherent optical transmission," *IEEE Journal of Lightwave Technology.*, Vol. 9, no. 2, 1991, pp. 182–193.

[30] Grosskopf, G., R. Ludwig, and H. G. Weber, "Crosstalk in optical amplifiers for two-channel transmission," *Electronics Letters*, Vol. 22, No. 17, 1986, pp. 900–902.

[31] Agrawal, G. P., "Amplifier-induced crosstalk in multichannel coherent lightwave systems," *Electronics Letters*, Vol. 23, 1987, pp. 1175–1177.

[32] Inoue, K., "Crosstalk and its power penalty in multichannel transmission due to gain saturation in a semiconductor laser amplifier," *IEEE Journal of Lightwave Technology.*, Vol. 7, No. 7, 1989, 1118–1124.

[33] Pettitt, M. J., A. Hadjifotiou, and R. A. Baker, "Crosstalk in erbium doped fiber amplifiers," *Electronics Letters*, Vol., No., 1989, pp.

[34] Rasmussen, T. et al., "Optimum placement of a fiber optical amplifier in high bit rate direct detection systems," *Applied Optics*, Vol. 30, No. 30, 1991, pp. 4376–4378.

[35] Naka, A., and S. Saito, "In-line amplifier transmission distance determined by self-phase modulation and group velocity dispersion," *IEEE Journal of Lightwave Technology*, Vol. 12, No. 2, 1994, pp. 280–287.

[36] Lin, C. and W. Way, "Optical amplifiers for multi-wavelength broadband distribution networks," *European Conference on optical Communication ECOC'91*, Paris, 1991, Proc. Vol. II, invited paper, pp. 125–133.

[37] Goldstein, E. L., et al., "Inhomogeneously broadened fiber-amplifier cascades for transparent multiwavelength lightwave networks," *IEEE Journal of Lightwave Technology.*, Vol. 13, No. 5, 1995, pp. 782–790.

[38] Tachibana, M., et al., "Erbium-doped fiber amplifier with flattened gain spectrum," *IEEE Photonics Technology Letters*, Vol. 3, 1991, p. 118.

[39] Inoue, K. et al., "Tunable gain equalization using mach-Zehnder optical filter in multistage amplifiers," *IEEE Photonics Technology Letters*, Vol. 3, 1991, pp. 718.

[40] Chraplyvy, A. R., J. A. Nagel, and R. W. Tkach, "Equalization in an amplified WDM lightwave transmission systems," *IEEE Photonics Technology Letters*, Vol. 4, No. 8, 1992, pp. 920–922.

[41] Willner, A. E. and S. M. Hwang, "Transmission of many WDM channels through a cascade of EDFA's in long-distance links and ring networks," *IEEE Journal of Lightwave Technology.*, Vol. 13, No. 5, 1995, pp. 802–816.

[42] Agrawal, G. P., *Fiber-Optic Communication Systems*, New York: John Wiley and Sons, 1992.

Chapter 4

Direct Detection Lightwave Systems

The function of an optical receiver is to detect and extract the information signal from the modulated optical wave. The information signal may be analog or digital, which is modulated by intensity modulation (IM) or external modulation schemes. The photodetectors such as an APD or a PIN diode are used for converting the incident modulated optical wave into an electrical signal. Then the electrical signal is processed by an amplifier and filter, resulting in the information signal. A direct-detection receiver is broadly classified into two categories: (1) the digital optical receiver or (2) the analog optical receiver. In the case of a digital optical receiver, the receiver recovers the information signal from the ON/OFF keying (OOK) modulated optical pulse stream, which is commonly employed in practical lightwave systems. An analog optical receiver, on the other hand, extracts the information signal from the AM, FM, or pulse modulated optical signals.

In recent years, the optical amplifiers have been used in conjunction with a direct detection receiver for obtaining higher receiver sensitivity and longer repeater spacing. As discussed in Chapter 3, an optical amplifier can be used as a preamplifier or an in-line amplifier in conjunction with analog or digital optical receivers. The use of optical amplifiers has opened up the possibility of new and more complex network structures as well as linking continents through oceans. The subject of optical amplifiers in direct detection systems is one of the most important developments in lightwave communication systems.

The analog optical receivers are discussed in Chapter 8 in the context of analog-modulated cable television (CATV) distribution systems. This chapter is devoted to the theory and the design of direct detection digital optical receivers with and without optical amplifiers. Section 4.1 describes the basic principle of operation of a direct detection digital optical receiver (without optical amplifiers), sensitivity analysis, and the receiver configurations. Section 4.2 describes the theory of direct detection systems with optical amplifiers along with the associated field trials and experiments. Finally, performance degradation issues are discussed in Section 4.3.

4.1 DIRECT DETECTION DIGITAL OPTICAL RECEIVER

A direct detection digital optical receiver looks for the presence or absence of light during the transmitted bit period. In this section, we discuss the theory of direct detection receivers without optical amplifiers [1–6]. The block diagram of a direct detection digital optical receiver is shown in Figure 4.1. It consists of a photodetector, which converts an optical signal into an electrical signal. The electrical signal from the photodetector output is amplified by a low-noise preamplifier, which is followed by an equalizer that optimizes the frequency response of the signal. Finally, the transmitted data is recovered by passing the equalized signal through a data detector. The data detector in a receiver measures the received signal during each pulse interval and determines whether a data bit 1 or 0 was sent during that interval. The data decision is accomplished by setting a threshold voltage between the two data bit levels and comparing the received voltage with the threshold voltage.

The photodetector (APD or PIN) is operated in a reverse-biased mode so that the incident photons are absorbed in a depleted semiconductor region containing a strong electric field. Due to the presence of the electric field, the electron-hole pairs are created by photon absorption, which in turn are separated and collected at the detector terminals. Hence, the resulting current is given by

$$i_s(t) = M\mathcal{R}p(t) \tag{4.1}$$

where M denotes the multiplication or gain in a detector and it is equal to unity for a PIN diode and $M > 1$ for an APD diode, and $\mathcal{R}$ denotes the responsivity of a photodetector.

The process of photocurrent generation is random in nature. The statistics of the shot noise generated in a p-i-n diode receiver follows the Poisson process, but in the case of an APD receiver, the shot noise generated is signal-dependent and is treated as doubly stochastic marked and filtered Poisson process (MFPP) [7]. To understand the basic mechanism of this random process from the statistical point of view, assume an incident optical pulse $p(t)$ that produces primary electrons. The number of primary electrons is a Poisson random variable that depends on the intensity of the optical signal. Each primary electron is then multiplied by a random

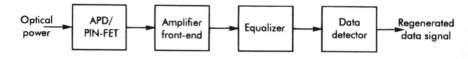

Figure 4.1 Block diagram of a direct detection digital optical receiver.

avalanche gain to produce a number of secondary electrons. The avalanche diode's output current, which consists of "bunches" of secondary electrons is then amplified along with the Gaussian thermal noise.

The equivalent circuit diagram of a typical digital optical receiver is shown in Figure 4.2. In the equivalent circuit diagram, $i_s(t)$ represents the photodetector current with capacitance C_d, $i_b(t)$ denotes the thermal noise current due to the detector bias resistor R_b, and the amplifier is represented by its input impedance (R_a in parallel with C_a) along with the gain A. The noise sources of an amplifier are represented by the input noise current source $i_a(t)$, which arises from the thermal noise of the amplifier input resistance, and the noise voltage source $e_a(t)$ denotes the thermal noise of the amplifier channel. These noise sources are assumed to be white Gaussian in statistics. The optical signal $p(t)$ incident on the photodetector can be described in terms of the binary digital pulse train as [6]

$$p(t) = \sum_{k=-\infty}^{\infty} b_k h_P(t - kT) \tag{4.2}$$

where $b_k = 1$ or 0, representing the kth message digit, $h_P(t)$ is the received pulse shape, assumed to be positive for all t, and T is the bit period. The received pulse $h_P(t)$ at the input of a photodetector is normalized to have a unit area, i.e., $\int_{-\infty}^{\infty} h_P(t)dt = 1$. The photodetector current is amplified by an amplifier and then filtered by an equalizer to yield the average voltage output as

$$\langle v_o(t) \rangle = AM\mathcal{R}p(t){*}h_B(t){*}h_{eq}(t) \tag{4.3}$$

where $h_{eq}(t)$ is the impulse response of an equalizer and the $*$ operator denotes the convolution. The term $h_B(t)$ represents the impulse response of the bias circuit, which can be found from the inverse Fourier transform of the transfer function. The transfer function of the bias circuit is given by [2]

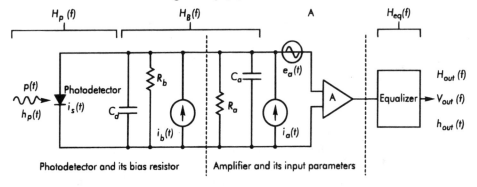

Figure 4.2 Equivalent circuit diagram of a typical digital optical receiver. (*Source:* [2]. Reprinted with the permission of McGraw-Hill.)

$$H_B(f) = \frac{R}{(1 + j\omega RC)} \tag{4.4}$$

where R represents the parallel combination of R_a and R_b, and C denotes the parallel combination of C_a and C_d. Using (4.2) in (4.3), the average output voltage from the equalizer can be written as

$$\langle v_o(t) \rangle = \sum_{k=-\infty}^{\infty} b_k h_o(t - kT) \tag{4.5}$$

where

$$h_o(t) = AM\mathcal{R}h_P(t) * h_B(t) * h_{eq}(t) \tag{4.6}$$

which defines the shape of the pulse at the equalizer output. The shape of the equalizer output is just the inverse Fourier transform of $H_o(f)$, which is given by

$$H_o(f) = AM\mathcal{R}H_P(f)H_B(f)H_{eq}(f) \tag{4.7}$$

where $H_P(f)$ is the Fourier transform of the received pulse shape $h_P(t)$ and $H_{eq}(f)$ is the transfer function of the equalizer. Hence, for a given input pulse shape and biasing circuitry, one can choose the equalized pulse shape that minimizes the errors caused by noise and intersymbol interference. The desired equalized output waveform is usually characterized by a raised-cosine pulse output [6]

$$h_o(t) = \frac{\sin \pi\tau}{\pi\tau}\left[\frac{\cos \pi\beta_p\tau}{1 - (2\beta_p\tau)^2}\right] \tag{4.8}$$

where $\tau = t/\pi$. The parameter β_p varies between 0 and 1, which represents the bandwidth of the pulse. The shape of the raised-cosine pulse waveform is shown in Figure 4.3 for different values of β_p. The actual voltage output at the equalizer can be written as

$$v_o(t) = \langle v_o(t) \rangle + v_N(t) \tag{4.9}$$

where $v_N(t)$ is the receiver noise voltage, which causes $v_o(t)$ to deviate from its average value. The receiver noise voltage comprises of the shot noise generated in the photodetector and the thermal noise in amplifier and resistors. For APD receivers, the shot noise due to the incident optical power is dominant in comparison with other noise sources. The PIN diode receivers, on the other hand, are usually limited by the thermal noise rather than the shot noise.

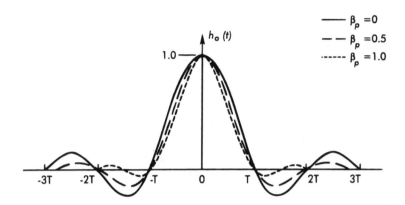

Figure 4.3 Plot of the raised cosine pulse waveform. (*Source:* [2]. Reprinted with the permission of McGraw-Hill.)

Next, let us look at the performance of the receiver. The performance of a receiver is commonly measured by the bit-error-rate (BER) and sensitivity. BER is defined as the ratio of the number of errors occurred during a specific time interval to the total number of data bits transmitted during that interval. The typical BER for lightwave communication systems is on the order of 10^{-9} to 10^{-12}. For example, BER $= 10^{-9}$ means that on the average, one error occurs for every billion data bits transmitted. The value of the BER mainly depends on the signal-to-noise at the equalizer output. The computation of BER requires the probability of distribution of the signal at the equalizer output. BER is usually computed by assuming equal probability of 1 and 0 occurrences as [8]

$$\text{BER} = \frac{1}{2}[P_1(V_T) + P_0(V_T)] \tag{4.10}$$

where V_T denotes the threshold voltage and the terms $P_1(V_T)$ and $P_0(V_T)$ represent the probability of errors when 1 or 0 are transmitted, which can be found from [8].

$$P_1(V_T) = \int_{-\infty}^{V_T} p(v/1) \, dv \tag{4.11a}$$

$$P_0(V_T) = \int_{V_T}^{\infty} p(v/0) \, dv \tag{4.11b}$$

where $p(v/1)$ is the probability that the output voltage is v when 1 is transmitted and similarly $p(v/0)$ denotes the probability of the output voltage v when 0 is sent.

These probabilities are also known as conditional probability distribution functions. The computation of $p(v/1)$ and $p(v/0)$ requires the statistics of the noise voltage at the time of sampling. In general, it is a difficult task to obtain the statistics of the output voltage due to the nonGaussian nature of the photodetection process, particularly for an APD receiver. However, the avalanche gain distribution has been well characterized in the literature [9–11].

Another important parameter that is associated with BER is the receiver sensitivity. The receiver sensitivity is defined as the minimum signal power required at the input of a receiver (in dBm or photons/bit) to achieve a preassigned BER. In the following subsections, the quantum-limited as well as practical receiver sensitivities are discussed.

4.1.1 Quantum-Limited Receiver Sensitivity

In optical receivers, even for the zero amplifier noise and zero dark current, the receiver sensitivity is limited by the quantum nature of the photodetection process (i.e., by the shot noise due to the incident optical power), which is called the quantum limit. It represents the best possible receiver sensitivity that can be achieved and is often used as a reference to compare practical receivers.

To estimate the quantum-limited receiver sensitivity, let us assume an ideal PIN diode having a unity quantum efficiency and zero dark current. Let us consider the two physical signals at the output of a photodetector, which depend only on the received optical power. These signals are represented over a transmitted bit period T as [3]

$$S_1(t) = \frac{N_p}{T} \quad \text{for a bit 1} \quad 0 \le t \le T \tag{4.12}$$

$$S_0(t) = 0 \quad \text{for a bit 0} \quad 0 \le t \le T \tag{4.13}$$

where N_p in (4.12) represents the average number of photons that are produced by an optical pulse during the bit period, i.e.,

$$N_p = \frac{\Re P_S T}{q} \tag{4.14}$$

where q is the electron charge and P_S is the signal power. The direct detection receiver integrates the signal over the bit period T or in otherwords, counts the number of received photons. In this case, an error occurs if an optical pulse is present at the receiver, with zero number of photons/bit. We need to consider Poisson statistics for the quantum-limited signal detection and N_p in this case denotes the mean of the Poisson process. Assume that there are no errors due to shot noise on the bit 0 and that the total error probability is mainly determined by the probability of the error occurred during a bit 1 period [3]

$$P_e = \frac{1}{2} e^{-N_p} \qquad (4.15)$$

If $P_e = 10^{-9}$, then from (4.15), one finds that $N_p = 20$ as the required minimum number of photons contained in an optical pulse, which is called the shot noise or quantum-limited receiver sensitivity.

4.1.2 Practical Receiver Sensitivity

In practical receivers, the computation of error probabilities require the probability distribution of v_o at the sampling times when the binary bits 1 and 0 are transmitted. The accurate calculation of these error probabilities is quite complicated. During the past several years, various methods have been developed to compute the error probabilities as well as the associated sensitivity of practical receivers [7,12–23]. These methods are broadly classified into five categories: (1) exact computation; and (2) computer simulation; (3) Chernoff bounds; (4) numerical methods; and (5) approximation methods. Several of these methods are compared in [12] in terms of the accuracy and computation time.

The first method is based on the exact probability distribution of the avalanche gain distribution, which serves as the standard for comparing the other methods. In the second method, the computer simulation based on Monte Carlo technique with importance sampling has been utilized. The third method involves the use of Chernoff equality to obtain the bounds on the error probabilities. The bound method is useful, since a complete description for the signal and noise processes can be characterized in terms of moment generating functions. However, much tighter bounds can be obtained by using the modified Chernoff inequality, called the Modified Chernoff bounds [13].

The fourth method uses the numerical techniques such as Gaussian quadrature rule, Gram-Charlier series, and method of moments [7,14–19]. The last method is based on saddlepoint [20] and Gaussian approximation [6] techniques. By using the saddlepoint approximation, one can accurately compute the error probabilities, but it would require time-consuming averaging over all possible signal sequences, if the effect of ISI has to be considered. On the other hand, the Gaussian approximation technique yields a closed form expressions for bit error probabilities with reasonable estimates and an insight for the receiver design. However, the Gaussian approximation technique underestimates the threshold setting and overestimates the optimal avalanche gain.

In this section, we state the results obtained by the Gaussian approximation technique without any derivations [21]. In this approximation, for a given input optical pulse, the output voltage $v_o(t)$ is assumed as a Gaussian random variable (also called signal-to-noise ratio approximation). Based on this approximation, one can calculate the BER with the knowledge of the mean and standard deviation of the output signal $v_o(t)$. To begin with the Gaussian approximation technique, let us assume that there is no optical power incident on the photodetector, when the bit 0 was sent (zero extinction). Let $h_P(t) = \delta(t)$, a unit impulse function for the optical pulse so that the individual received pulses do not overlap (no intersymbol interference) at the sampling times $t_S = kT$ and the maximum value of $h_o(t_S = 0) = 1$ and $h_o(t_S = kT) = 0$ for $k \neq 0$. Hence, the conditional error probabilities $p(v/1)$ and $p(v/0)$ are approximated with Gaussian probability density functions. The corresponding conditional error probabilities are given by [8]

$$p(v/1) = \frac{1}{\sqrt{2\pi\sigma_1^2}} \int_{-\infty}^{V_T} \exp\left[-\frac{(v - \mu_1)^2}{2\sigma_1^2}\right] \tag{4.16}$$

$$p(v/0) = \frac{1}{\sqrt{2\pi\sigma_0^2}} \int_{V_T}^{\infty} \exp\left[-\frac{(v - \mu_0)^2}{2\sigma_0^2}\right] \tag{4.17}$$

where μ_0 and σ_0^2 are the mean and the variance of the signal b_0. Similarly the parameters μ_1 and σ_0^2 represent the mean and the variance of the signal b_1. Using the change of variables

$$Q = \frac{(V_T - \mu_0)}{\sigma_0} = \frac{(\mu_1 - V_T)}{\sigma_1} \tag{4.18}$$

and by changing the limits of integration in (4.16) and (4.17), we get

$$p(v/1) = p(v/0) = \frac{1}{2}\operatorname{erfc}\left(\frac{Q}{\sqrt{2}}\right) \tag{4.19}$$

where erfc stands for the complimentary error function, given by [8]

$$\operatorname{erfc}(x) = \frac{2}{\sqrt{\pi}} \int_{x}^{\infty} \exp(-y^2) \, dy \tag{4.20}$$

Using (4.19) in (4.10) we get a closed form expression for the BER as

$$\mathrm{BER} = \frac{1}{2}\operatorname{erfc}\left(\frac{Q}{\sqrt{2}}\right) \tag{4.21}$$

From (4.18), the parameter Q can be obtained in terms of the values of the mean and variance as

$$Q = \frac{\mu_1 - \mu_0}{\sigma_1 + \sigma_0} \tag{4.22}$$

Q is simply the ratio of the average signal level and the r.m.s noise at the threshold detector input. The required value of Q is approximately equal to 6 for a BER = 10^{-9}. Following the analysis of Smith and Garett [21], the sensitivity of a practical receiver can be expressed as

$$P_{APD} = \frac{hcBQ}{2\lambda\eta} \left[QI_2 F(M) + \frac{2}{M} \sqrt{Z + \frac{I_2}{QB} I_{DM} M_D^2 F_D} \right] \tag{4.23}$$

where h is Planck's constant, c is the velocity of free space, B is the bit rate, M is the gain, η is the photodetector efficiency, and $F(M)$ is the excess noise factor of the APD. The parameter I_{DM} represents the multiplied dark current characterized by the noise and gain parameters F_D and M_D, respectively. The term Z in (4.23) defines the thermal noise characteristic of the receiver amplifier, which is given as [5]

$$Z = \frac{I_2}{q^2 B} \left[S_1 + \frac{S_E}{R^2} \right] + S_E B I_3 \frac{(2\pi C)^2}{q^2} \tag{4.24}$$

where q is the electronic charge; R and C are the equivalent circuit parameters; and I_2 and I_3 are the weighting functions, which depend only on the input optical pulse shape. For NRZ coding format and equalized raised cosine output pulse, $I_2 = 0.562$, $I_3 = 0.0868$. Similarly for RZ code with 50% duty cycle the values of I_2 and I_3 are 0.403 and 0.0361, respectively. The expressions S_E and S_I denote the spectral densities of the amplifier noise voltage and current sources, respectively. Their expressions are given in Section 4.1.2.

For the case of PIN diode receivers, we have $M = 1$, $F(M) = 1$, and $I_{DM} = 0$, which reduces (4.23) into

$$P_{PIN} = \frac{hcBQ}{2\lambda\eta} (QI_2 + 2\sqrt{Z}) \tag{4.25}$$

It can be seen from (4.25), that the sensitivity of a PIN diode receiver is primarily limited by the thermal noise characteristic parameter Z. When $QI_2 < 2\sqrt{Z}$, the thermal noise limited receiver sensitivity is $hcBQ\sqrt{Z}/\lambda\eta$. The main goal in a receiver design is to choose the bias circuit resistor and the response of the equalizer so that a receiver that samples $v_{out}(t)$ at sampling times $t_S = kT$ can make decisions as to which value of b_k has been assumed with minimum probability of error.

4.1.3 Receiver Configurations

An optical receiver can be broadly classified into three types: (1) the low impedance receiver; (2) the high impedance receiver; or (3) the transimpedance receiver. The noise characteristics of each of the above receiver configurations depend on the front-end amplifying device such as a bipolar junction transistor or a field effect transistor [24–28]. The following subsections describe the noise characteristics of these configurations in terms of the spectral densities of the respective noise sources by assuming a FET front-end amplifier.

4.1.3.1 Low-Impedance Receiver

The schematic of a low impedance receiver configuration is shown in Figure 4.4. This receiver configuration offers widest dynamic range and bandwidth with lowest sensitivity. It utilizes a standard 50 Ohm termination as a biasing resistor and a coaxial line at the front-end to suppress any standing waves (reflections) for obtaining uniform frequency response. In this configuration, the noise sources are characterized by the resistor noise and the amplifier noise. The resistor noise—also called the thermal noise—is specified by its spectral density as [2]

$$S_I = \frac{4k_B T_k}{R_b} \qquad \text{A}^2/\text{Hz} \qquad (4.26)$$

where k_B is the Boltzman's constant, T_k is the temperature in Kelvin, and R_b is the

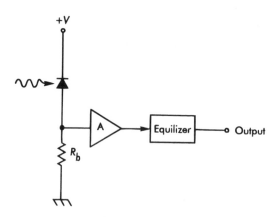

Figure 4.4 Low or high impedance receiver configuration. Here, R_b is the biasing resistor.

bias circuit resistance. The FET device noise, due to the thermal noise of the channel resistance, is characterized by the its voltage noise spectral density as [2]

$$S_E = \frac{8k_B T_k}{3g_m} \quad \text{V}^2/\text{Hz} \tag{4.27}$$

where g_m is the transconductance of the FET device.

4.1.3.2 High-Impedance Receiver

The high impedance receiver configuration is same as that of low-impedance circuit except for the high resistance value for the biasing resistor. The high impedance receiver circuit offers the lowest noise level and hence the highest receiver sensitivity. However, this configuration has limited dynamic range due to the high-input bias resistance. Because of the high load resistance at the front-end, the frequency response of the receiver is limited by the RC time constant. Due to this large time constant, the high impedance receiver tends to integrate the detected signal and is often referred to as the integrated front-design. The high-impedance receiver usually requires an equalizer following the front-end amplifier to improve the receiver frequency response. The thermal noise (S_E) due to the amplifier input resistance can be neglected, because of the higher input impedance of this configuration. The voltage noise spectral density is given by (4.27).

4.1.3.3 Transimpedance Receiver

Figure 4.5 shows the schematic diagram of a transimpedance receiver configuration. This configuration offers a good dynamic range and is designed to take advantage of the negative feedback effect for achieving higher bandwidth. In this case, an equalizer circuit may not be required. However, this receiver has a higher noise level due to the thermal noise of the feedback resistor and hence offers lower receiver

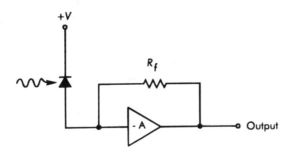

Figure 4.5 Transimpedance receiver configuration.

sensitivity. The main noise source is in this case is the thermal noise, which can be characterized by replacing R_b and R by

$$R'_b = \frac{R_b R_f}{R_a + R_b} \tag{4.28}$$

$$R' = \frac{R'_b R_a}{R'_b R_a} \tag{4.29}$$

where R_f is the feedback resistance. In principle, the receiver sensitivity degradation can be minimized by choosing the largest value for the feedback resistance, so that its thermal-noise contribution to the receiver noise is negligible.

4.1.4 Comparison of Receiver Sensitivity

In this section, the sensitivity results of APD and PIN diode receivers are discussed. In addition, the results pertaining to experimental, commercial, and state-of-the-art receivers are compared. The sensitivity results of Si APD/PIN diode, InGaAs PIN/APD diode, and Ge APD receivers are shown in Figure 4.6 (a,b) as a function of the bit rate. From this figure, it can be seen that the APD receivers offer better receiver sensitivity than the PIN diode receivers, due to the internal gain of an APD. However, the APD receiver is relatively expensive and requires a very high reverse bias voltage. Additionally, the APD receivers offer lower bandwidths than the PIN diode receivers because of the longer transition time required due to the avalanche process. From Figure 4.6(a), it can be seen that when a PIN diode is used, a low noise front-end amplifier is necessary to achieve high sensitivity. The combination of low noise transimpedance front-end amplifier and a Si APD offers the best sensitivity as well as wide dynamic range for short wavelength systems.

From Figure 4.6(b), it can be seen that as similar to short wavelength system, the Ge and InGaAs APD receivers offer better sensitivity than the PIN diode receivers, particularly at higher bit rates. However, for high bit rate applications, (over 1 Gbps) the sensitivity improvement obtained by the Ge and InGaAs APD receivers is limited by the lower gain-bandwidth product. The PIN diode receiver requires a low noise front end amplifier to achieve the best receiver sensitivity. The high-impedance FET front end amplifier is normally chosen to obtain the best sensitivity.

In recent years, the sensitivity of PIN diode receivers is significantly improved by reducing the noise of front-end amplifiers. The front-end amplifier noise is minimized by using a hybrid integrated PIN diode and FET amplifier (called PINFET) having very low input capacitance and leakage currents [29,30]. Comparable performance to an APD receiver can be obtained by choosing the bias resistor or the

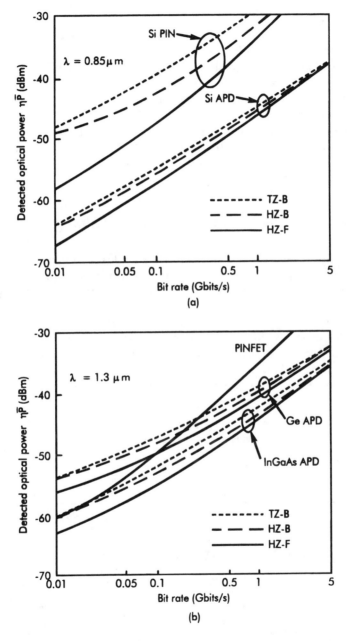

Figure 4.6 Sensitivity of optical receivers at (a) 0.85 μm and (b) 1.3 μm. Here, HZ and TZ represent the high impedance and transimpedance receiver configurations with bipolar (B) and FET (F) front end-amplifiers. (*Source:* [4]. © 1984 IEEE. Reprinted with permission.)

feedback resistor of the transimpedance configuration sufficiently large so that the thermal noise can be negligible. In addition, high-speed optical receivers for multigigabit applications (>10 Gbps) have been developed by monolithically integrating the PIN diode with high electron mobility transistors (HEMTs) and HBTs (heterojunction bipolar transistors)][31–34]. These optoelectronic integrated circuits are discussed in Chapter 7. Typical sensitivities pertaining to experimental and state-of-the-art direct detection receivers are shown in Table 4.1.

The commercial direct detection receivers offer a sensitivity of about −33 dBm in the 0.85 μm window at a bit rate of 150 Mbps [41,42]. Similarly, in the 1.3–1.55 μm region, the receiver sensitivity is in the range of −33 to −38 dBm over a bit rate of 200–300 Mbps [43,44].

4.2 DIRECT DETECTION WITH OPTICAL AMPLIFIERS

The use of a traditional repeater to compensate the loss in a lightwave link is quite cumbersome due to the photon-electron conversion, electrical amplification, retiming, pulse shaping, and electron-photon conversion. Therefore, the direct amplification of the light signal would be beneficial in system applications such as long-haul, LANs, and WDM networks, employing direct detection receivers.

In this section, we discuss the use of an optical amplifier as a preamplifier or as in-line amplifier in direct detection receiving systems. Section 4.2.1 discusses the receiver structure, and Section 4.2.2 outlines the BER and sensitivity computations. Finally, Section 4.2.3 reviews the recent field trials and experiments.

Table 4.1
Sensitivity of Direct Detection Receivers

Wavelength h(μm)	Bit rate (Mbps)	Receiver	Sensitivity (dBm)	Format	Reference
0.85	50	Si APD FET-HZ	−56.6	RZ	[35]
	1.5	Si PIN FET-TZ	−57.2	NRZ	[36]
1.3	8000	InGaAs APD FET-HZ	−25.8	NRZ	[37]
	565	InGaAs PIN FET-HZ	−38.5	NRZ	[38]
1.51	4000	InGaAs APD FET-HZ	−31.2	NRZ	[39]
1.52	140	Ge APD FET-TZ	−49.3	NRZ	[40]
1.53	1200	InGaAs PIN FET-HZ	−36.5	NRZ	[29]
1.55	10	PIN/HEMT OEIC	−28.0	NRZ	[31]
1.55	20	PIN/HBT OEIC	−17.0	NRZ	[34]

Note: HZ: High impedance, TZ: Transimpedance, RZ: Return to zero, NRZ: Non return to zero; HEMT: High mobility electron transistor; HBT: Heterojunction bipolar transistor; OEIC: optoelectronic integrated circuit.

4.2.1 Receiver Structure

The schematic diagram of a direct detection receiver with an optical preamplifier is shown in Figure 4.7. The incoming signal is amplified with a frequency flat gain G, and the desired signal is passed through a polarization filter. (It blocks all the light in orthogonal polarization.) Then the signal is processed by an optical bandpass filter before the photodetection and the current output from the photodetector is passed through an integrate and dump circuit (post detection filter) for data detection.

Assuming an ASK modulation with NRZ format, the electric signal at the input of an amplifier is given by [45]

$$E_i(t) = \sqrt{2P_{in}} \cos[2\pi\nu t + \phi(t)] \qquad (4.30)$$

where ν is the optical frequency, P_{in} is the amplifier input power, and $\phi(t)$ is the laser phase noise. The phase noise broadens the signal spectrum and may require a higher optical bandwidth. The larger optical bandwidth allows more spontaneous emission noise from the amplifier and hence degrades the receiver performance. The signal at the output of an amplifier is given as

$$E_o(t) = \sqrt{2P_S} \cos[2\pi\nu t + \phi(t)] + n_{SP}(t) \qquad (4.31)$$

where $n_{SP}(t)$ represents the amplified spontaneous emission noise, assumed to be Gaussian with zero mean. The term $P_S = (C_i C_o L P_{in})$ represents the received signal power at the input of a receiver, where C_i denotes the input coupling efficiency, C_o is the output coupling efficiency, and L is the loss between the amplifier and the receiver. The filtered optical signal is detected by a photodetector (PIN diode), yielding an output current [45]

$$i(t) = \Re|E_F(t)|^2 + n_S(t) + n_{TH}(t) \qquad (4.32)$$

where $E_F(t)$ is the complex amplitude of the filtered optical signal, which contains

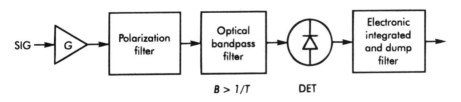

Figure 4.7 Direct detection lightwave receiver with an optical preamplifier. (*Source:* [45]. © 1991 IEEE. Reprinted with permission.)

the filtered amplifier spontaneous emission noise. The term $n_S(t)$ denotes the photodetector shot noise, and $n_{TH}(t)$ represents the thermal noise of the receiver.

4.2.2 Sensitivity Analysis

Equation (4.32) is a well studied problem in communication theory, which is a case of square law detection of digital signals in Gaussian noise. The addition of a photodetector leads to nonGaussian statistics for which a closed form expression for BER and sensitivity is difficult to obtain. Several approaches have been used to calculate the sensitivity of a direct detection lightwave receiver with an optical preamplifier [45–54]. In one approach [45], a detailed sensitivity analysis of the receiver is carried out taking into account the amplifier spontaneous noise and the laser linewidth, assuming a high gain amplifier with negligible shot noise and thermal noise contributions.

In another approach [49], the highest sensitivity analysis of the receiver is calculated for a BER = 10^{-9}, taking into account the profile of the amplifier response by assuming zero linewidths and a unity spontaneous emission factor. In this case, the sensitivity of the amplified receiver is given by

$$N_p = 36 + 6\sqrt{B_oT} \tag{4.33}$$

where N_p is the number of photons per bit, T is the bit period, and $B_o = kB_n$, is the bandwidth of the amplifier, where k is a constant that depends on the response characteristics of the amplifier, and B_n represents the noise bandwidth of the amplifier. The parameter $k = 1$, 0.707, and 0.5, for rectangular, Gaussian, and exponential shaped impulse responses, respectively. If the B_oT product equals 1 (i.e., $B_o = 1/T$), then the number of photons per bit required is only 42. It can be seen from (4.33), the receiver sensitivity degrades with an increase in the optical bandwidth.

To obtain a closed form expression for the BER, we assume Gaussian approximation for the signal and noise components at the time of sampling with negligible laser linewidths [48]. First, consider the received signal power after square law detection in the receiver, which is given by [50]

$$S = (G\Re P_S)^2 \tag{4.34}$$

where G is the amplifier gain, $\Re$ denotes the responsivity of the photodetector, and P_S denotes the received signal power. Next, the total average noise power resulting at the output of a photodetector is given by

$$N_T = N_S + N_{S-SP} + N_{SP-SP} + N_{TH} \tag{4.35}$$

where N_S is the shot noise, N_{S-SP} is the beat noise component between the signal and the spontaneous noise, $N_{SP\text{-}SP}$ is the spontaneous-spontaneous beat noise, and N_{TH} is the thermal noise of the receiver. It is important to note that $N_{SP\text{-}SP}$ is a dominant factor when the optical bandwidth is large, otherwise the receiver is primarily limited by $N_{S\text{-}SP}$. The noise variance of these components can be expressed as [50]

$$\sigma_S^2 = 2q\Re L[GP_{\text{in}}C_iC_o + n_{SP}(G - 1)h\nu B_o C_o]B_e \tag{4.36}$$

$$\sigma_{S-SP}^2 = 4\Re q\eta C_i C_o^2 L^2 P_{\text{in}} G(G - 1)n_{SP}B_e \tag{4.37}$$

$$\sigma_{SP-SP}^2 = (\eta q C_o L)^2[n_{SP}(G - 1)]^2(2B_o B_e - B_e)^2 \tag{4.38}$$

$$\sigma_{TH}^2 = I_{TH}^2 B_e \tag{4.39}$$

where C_i denotes the input coupling efficiency, C_o is the output coupling efficiency, $\Re$ is the responsivity of the photodetector, P_{in} is the amplifier input power, and L is the loss between the amplifier and the receiver as defined earlier. Other terms include η as the quantum efficiency of the photodetector, q is the electronic charge, $h\nu$ is the photon energy, and n_{SP} is the spontaneous emission factor of the amplifier. The terms B_e and B_o represent the electrical and optical bandwidths respectively. The factor I_{TH}^2 represents the equivalent thermal noise current of the receiver. If an APD is used as a photodetector, an avalanche multiplication factor M and an excess factor $F(M)$ should be considered in (4.36)–(4.39).

Now the BER can be computed by using (4.21) and (4.22). The mean values for (4.22) are given by

$$\mu_1 = \sqrt{S(1)} \tag{4.40}$$

$$\mu_0 = \sqrt{S(0)} \tag{4.41}$$

where $S(1) = G\Re P_S$ and $S(0) = 0$ (by assuming a very large value of extinction ratio). Similarly, the terms σ_1 and σ_2 for (4.22) can be found from

$$\sigma_1 = \sqrt{(\sigma_S^2 + \sigma_{S-SP}^2 + \sigma_{SP-SP}^2 + \sigma_{TH}^2)} \tag{4.42}$$

$$\sigma_0 = \sqrt{(\sigma_S^2 + \sigma_{SP-SP}^2 + \sigma_{TH}^2)} \tag{4.43}$$

Next let us discuss the performance of a direct detection receiver using optical amplifiers. As mentioned before, the two main applications of an optical amplifier are: (1) the preamplifier and (2) the in-line amplifier. When it is used as a preamplifier, the optical amplifier boosts the optical signal prior to the photodetector and hence increases the receiver sensitivity. In in-line amplifier applications, the amplifier operates as a repeater and provides the required gain for compensating the loss between the transmitter and receiver. Additionally, several in-line amplifiers can be cascaded to increase the system span.

First let us discuss the application of an optical amplifier as a preamplifier. Figure 4.8 shows the sensitivity of a 1.55 μm receiver versus amplifier gain for different optical filter bandwidths (assuming rectangular shape amplifier response). From Figure 4.8, it can be seen that the receiver achieves the highest sensitivity at lower optical bandwidths. The best receiver sensitivity is achieved when $B_o = 2B_e$, which is approximately equal to -39.4 dBm (181 photons/bit). Higher optical bandwidth means more spontaneous emission noise collected and hence a decrease in receiver sensitivity. The decrease in receiver sensitivity with an increase in optical bandwidth is often measured by the power penalty. The power penalty is larger for higher gains when the receiver is closer to the best sensitivity level. For higher gains, the spontaneous-spontaneous noise and signal-spontaneous noise become dominant. At lower amplifier gains, the receiver is limited by the thermal noise, and consequently the sensitivity of the receiver improves 1 dB for every dB of gain. The above results were obtained by assuming an infinite extinction ratio. In practice, an extinction ratio of 20:1 or better incurs a power penalty of less than 1 dB [48].

The effect of laser linewidths on the receiver sensitivity was carried out in [45], and the results reveal a similar effect, but with an accurate measure for the filter

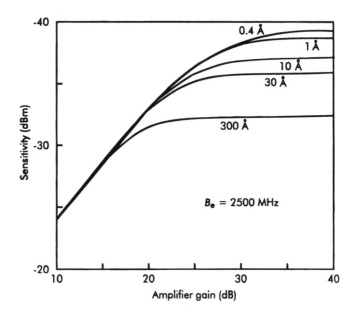

Figure 4.8 Receiver sensitivity versus amplifier gain for different values of optical bandwidths. The parameters used in the calculation are: B_e = 2.5 GHz, C_i = 0.31, C_o = 0.26, and n_{sp} = 1.4. (*Source:* [48]. © 1989 IEEE. Reprinted with permission.)

bandwidth and the power penalty. However, it is important to note that in the absence of an optical bandpass filter, laser phase noise is not a source of degradation [45].

Next, let us consider the application of the optical amplifier as an in-line amplifier. Figure 4.9 shows the BER curves as a function of received power for a system with an in-line amplifier for different amplifier input powers. From Figure 4.9, one can conclude that a negligible power penalty or a power penalty of less than 1 dB results, provided the amplifier input power is sufficiently large. Also, it can be seen that for power penalties less than 1 dB, no error floor exists for the BER, however in the case of a 2 dB penalty or above, an error floor for the BER exists.

Next, let us look at a system with cascaded in-line amplifiers. It is assumed that the gain of each amplifier exactly equals the loss between the two amplifiers. In this case, the cumulative effect of the amplifier noise can be obtained by replacing n_{SP} in (4.36)–(4.39) by $N_A n_{SP}$, where N_A is the number of amplifiers. Figure 4.10 shows the receiver sensitivity as a function of the number of amplifiers. From Figure 4.10, it can be seen that the maximum number of in-line amplifiers are 28, 280, and 2800 for the maximum input powers of -30, -20, and -10 dBm respectively with a maximum power penalty of 1 dB. The maximum number of amplifiers depends

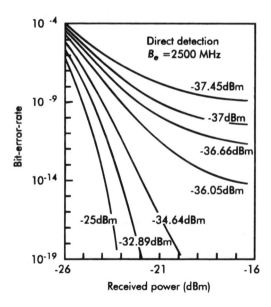

Figure 4.9 BER curves for direct detection in-line amplifier system with amplifier input powers of -25, -32.89, -34.64, -36.05, -36.66, -37, and -37.45 dBm, giving power penalties of 0, 0.5, 1.0, 2.0, 3.0, and 4 dB for BER = 10^9, respectively. (*Source:* [48]. © 1989 IEEE. Reprinted with permission.)

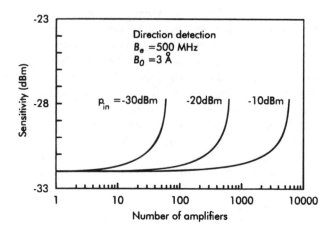

Figure 4.10 Direct detection receiver sensitivity versus number of in-line amplifiers. The parameters used in the calculation are: $B_e = 500$ MHz, $C_i = 1$, $G = 25$ dB, $\lambda = 1.32$ μm, $= n_{SP} = 1.4$. (*Source:* [48]. © 1989 IEEE. Reprinted with permission.)

on the bit rate, optical bandwidth, and amplifier input power and is limited by the accumulation of the amplifier noise along the link.

4.2.3 Field Trials and Experiments

In this section, we will outline the recent field trials as well as experiments of 1.55-μm direct detection lightwave systems with optical amplifiers. Table 4.2 shows some of the recent field trials and experiments.

Table 4.2
List of Field Trials and Experiments

Bit rate (Gbps)	Distance (km)	Fiber	In-line Amplifiers	Year	Reference
1.2	904	DSF	12 EDFA	1989	[55]
10	1500	DSF	22 EDFA	1991	[56]
10	9000	DSF	274 EDFA	1993	[57]
10	204	DST	1 EDFA	1993	[58]
10	4500	DSF	138 EDFA	1992	[59]

4.3 PERFORMANCE DEGRADATION ISSUES

In the previous sections, we have considered the amplifier noise and receiver noise as the main factors in determining the highest receiver sensitivity. In practice, other sources which degrade the receiver sensitivity must be taken into account [4,60–67]. First, the relative intensity noise (RIN) of the laser source can significantly degrade the SNR of analog optical receivers. However, the effect of RIN is negligible for digital optical receivers. Second, the effect of dark current in photodetectors can degrade the receiver sensitivity. For PINFET receivers, with dark currents of 10 nA, the sensitivity penalty is less than 0.5 dB for bit rates above 10 Mbps and less than 0.1 dB above 100 Mbps [4]. In the case of APDs, the effect of dark current can be a serious limiting factor, because of the avalanche multiplication of the current, particularly at high temperatures. For long wavelength APDs, assuming $k_i = 1$, the maximum allowable multiplied dark current to incur a sensitivity penalty of 0.5 dB is about 0.1 nA at 10 Mb/s and it is 10 nA at 1 Gbps [4].

Mode partition noise arises because of the random fluctuations in the partition of laser power among different longitudinal modes [62,63]. For long and/or dispersive transmission fiber lengths, the SNR can be degraded, which also limits the BER. The mode partition noise can be reduced or eliminated by operating the system at the minimum dispersive wavelength or by using stabilized single frequency lasers.

Fiber dispersion is a major factor in limiting the receiver sensitivity, particularly for systems using multimode fibers or LEDs [60,61]. Dispersion shifted fiber or dispersion compensation techniques can be utilized to compensate the chromatic dispersion of light sources. The effect of optical feedback to the laser cavity from splices, connectors, and other imperfections can also degrade the receiver sensitivity [64]. The degradation due to optical feedback can be minimized by using optical isolators. Another serious problem is the frequency chirp associated with the directly modulation of the laser current [65,66]. This phenomenon has been studied in [66] for a 4 Gbps signal for different DL products, where D is the fiber dispersion, and L is the transmission distance to keep the sensitivity penalty below 1 dB. The DL product should be less than or equal to 1.4 ns/nm. This means that for a dispersion of 0.1 ps/km/nm, the total transmission length for a 4 Gbps signal should be kept below 14,000 km.

The effect of nonzero extinction of the optical pulse on the receiver sensitivity, when a bit 0 was sent, has been analyzed in [23]. For practical systems, it is desirable to limit the extinction ratio to less than 0.1 dB so that the sensitivity degradation is within 1 or 2 dB. Additionally, the calculation of receiver sensitivity was based on the assumption that the signal is sampled at the peak of the voltage pulse. However, in practice, the decision instant is determined by the clock recovery circuit. Due to the noisy nature of the input to the clock-recovery circuit, the sampling time varies around its average value. The variations in sampling times is called the timing jitter. Due to this timing jitter, the SNR output is reduced and hence results in sensitivity degradation. It is shown in [67] that the penalty is negligible for $\sigma_j < 0.004$, where σ_j is the rms value of the jitter. Assuming a Gaussian distribution for the jitter signal with $\sigma_j \approx 0.07$ results in a 2 dB power penalty.

Problems

Problem 4.1

The received optical pulse at 0.85 μm has a rectangular shape with duration of 2 ns. Assume the peak power level of the pulse as 100 mW. Find the average number of electron-hole pairs produced by the received pulse, if the quantum efficiency of the photodetector is 0.8.

Problem 4.2

What is the probability that exactly 100 electron-hole pairs are generated from the received pulse in Problem 4.1. Assume the Poisson distribution for the input signal.

Problem 4.3

Calculate Z, the thermal noise characteristic for the following receivers at 50 Mbps, assuming a NRZ code with raised cosine pulse output.

  (a) High-impedance FET receiver with $g_m = 6$ mS, $R_a = 10$ M Ohm, $R_b = 1$ M Ohm, $C = 12$ pF, and $T_k = 300$ K.
  (b) Transimpedance FET receiver with $g_m = 6$ mS, $R_a = 10$ M Ohm, $R_b = 1$ M Ohm, $R_f = 1$ M Ohm, $C = 20$ pF and $T_k = 300$ K.

Problem 4.4

Calculate the sensitivity of APD receivers specified in Problem 4.3 to operate at 1.55 μm to achieve a BER $= 10^9$. Assume an APD with $k_i = 0.6$, $I_{DM} = 10$ pA, $M_D = 2$, $F_D = 0.8$ and $M = 100$. The quantum efficiency of an APD is 0.8.

Problem 4.5

Repeat Problem 4.4 assuming a PIN diode having a quantum efficiency of 0.9.

Problem 4.6

Find the sensitivity of a preamplified receiver in terms of photons/bit, assuming a Gaussian response for the amplifier. Assume $B_n T = 2.0$ and the operating bit rate as 1 Gbps.

Problem 4.7

Plot the BER versus received signal power of a preamplified 1.55-μm direct detection receiver, given the following parameters: $G = 25$ dB, $L = 1.0$, $\mathcal{R} = 0.9$, $n_{SP} = 1.5$, $B_e = 2 \cdot 5$ GHz, $B_o = 200$ Angstroms, $C_i = 0.25$, $C_o = 0.5$, $I_2 = 10$ pA/Hz. Assume $P_{in} = -30$ dBm.

Problem 4.8

In Problem 4.7, find the receiver sensitivity to achieve a BER $= 10^9$. Also estimate the BER floor.

Problem 4.9

Repeat Problem 4.7 for a cascaded system having 50 in-line amplifiers. Assume identical amplifiers, each having a gain of 25 dB.

Problem 4.10

Calculate the number of in-line amplifiers required for a system span of 200 km with a fiber loss of 0.2 dB/km. Assume identical amplifiers, each having a gain of 25 dB.

References

[1] Miller, S. E., and I. P. Kaminow, "*Optical Fiber Communications II*, Academic Press, New York, 1988.
[2] Keiser, G. E., *Optical Fiber Communications*, McGraw-Hill, New York, 1983.
[3] Jacobsen, G., *Noise in Digital Transmission Systems*, ARTECH House Inc., Norwood, MA, 1994.
[4] Muoi, T. V., "Receiver design for high-speed optical-fiber systems," *IEEE Journal of Lightwave Technology*, Vol. LT-3, No. 3, 1984, pp. 243–267.
[5] Brain, M. and Lee, T. P., "Optical receivers for lightwave communication systems," *IEEE Journal of Lightwave Technology*, Vol. LT-3, No. 6, 1985, pp. 1281–1300.
[6] Personick, S. D., "Receiver design for digital fiber optic communication systems," *Bell System Technical Journal*, Vol. 55, 1973, pp. 843–866.
[7] Cariolaro, G. C., "Error probability in digital fiber optic communication systems," *IEEE Transactions on Information Theory*, Vol. IT-24, No. 2, 1978, pp. 213–221.
[8] Ziemer, R. E. and W. H. Tranter, *Principles of communications*, Houghton Mifflin Co. Boston, 1990.
[9] McIntyre, R. J., "The distribution of gains in uniformly multiplying avalanche photodiodes: Theory," *IEEE Transactions on Electron Devices*, Vol. 19, 1972, pp. 703–712.
[10] Personick, S. D., "Statistics of a general class of avalanche detectors with applications to optical communications," *Bell System Technical Journal*, Vol. 50, 1971, pp. 3075–3095.
[11] Balaban, P., P. E. Fleischer, and H. Zucker, "The probability distribution of gains in avalanche photodiodes," *IEEE Transactions on Electron Devices*, 1976, pp. 1189–1190.

[12] Personick, S. D., P. Balaban, J. H. Bobsin, and P. R. Kumar, "A detailed comparison of four approaches to the calculation of the sensitivity of optical fiber system receivers," *IEEE Transactions on Communications*, Vol. COM-25, 1977, pp. 541–548.

[13] Da Rocha, J. R. F., and J. J. O'Reilly, "Modified Chernoff bound for binary optical communication," *Electronics Letters*, Vol. 18, No. 16, 1982, pp. 707–710.

[14] Hauk, W., F. Bross, and M. Ottka, "The calculation of error rates for optical fiber systems," *IEEE Transactions on Communications*, Vol. COM-26, No. 7, 1978, pp. 1119–1126.

[15] Dogliotti, R., A. Luvison, and G. Pirani, "Error probability in optical fiber transmission systems," *IEEE Transactions on Information Theory*, Vol. IT-25, No. 2, 1979, pp. 170–178.

[16] Rugemalira, R. A., "The calculation of average error probability in a digital fiber optical communication system," *Optical and Quantum Electronics*, Vol. 12, 1980, pp. 131–141.

[17] Mansuripur, M., J. W. Goodman, E. G. Rawson, and R. E. Norton, "Fiber optics receiver error rate prediction using the Gram-Charlier series," *IEEE Transactions on Communications*, Vol. COM-28, No. 3, 1980, pp. 402–407.

[18] O'Reilly, J. J. and Jose, R. F. daRocha, "Improved error probability evaluation methods for direct detection optical communication systems," *IEEE Transactions on Information Theory*, Vol. IT-33, No. 6, 1987, pp. 839–848.

[19] Cartledge, J. C., "A maximum entropy method for lightwave system performance evaluation," *IEEE Journal of Lightwave Technology*, Vol. LT-5, No. 11, 1987, pp. 1613–1617.

[20] Helstrom, C. W., "Computing the performance of optical receivers with avalanche diode detectors," *IEEE Transactions on Communications*, Vol. 36, No. 1, 1988, pp. 61–66.

[21] Smith, D. R., and I. Garrett, "A simplified approach to digital optical receiver design," *Opt. Quantum Electron.*, Vol. 10, 1978, pp. 211–218.

[22] Smith, R. G. and S. D. Personick, "Receiver design for optical fiber communication systems," in *Semiconductor Devices for Optical Communications*, H. Kressel, (Ed.), Springer-Verlag, New York, 1980.

[23] Hooper, R. C., and R. B. White, "Digital optical receiver design for non-zero extinction ratio using a simplified approach," *Opt. Quantum Electron.*, Vol. 10, 1978, pp. 279–282.

[24] Goell, J. E., "An optical repeater with high impedance input amplifier," *Bell Sys. Tech. Journal*, Vol. 53, 1974, pp. 629–643.

[25] Witkowicz, T., "Design of low noise fiber optic receiver amplifiers using J-FETs, *IEEE Journal on Solid State Circuits*, Vol. SC-13, 1978, pp. 195–197.

[26] Hullet, J. L., and Muoi, T. V., "A feedback receiver amplifier for optical transmission systems," *IEEE Transactions on Communications*, Vol. COM-24, 1976, pp. 1180–1185.

[27] Hullet, J. L., and S. Moustakas, "Optimum transimpedance broadband optical preamplifier design," *Optical and Quantum Electronics*, Vol. 13, 1981, pp. 65–69.

[28] Moustakas S., and J. L. Hullet, "Comparison of BJT and MESFET front ends in broadband optical transimpedance pre-amplifiers," *Optical and Quantum Electronics*, Vol. 14, 1982, pp. 57–60.

[29] Brian, M. P. P. Smyth, D. R. Smith, B. R. White, and P. J. Chidgey, "PINFET hybrid optical receivers for 1.2 Gb/s transmission systems at 1.3 and 1.55 μm wavelength, *Electronics Letters*, Vol. 20, 1984, pp. 894–896.

[30] Smith, D. R., et al., "PINFET hybrid optical receiver for 1.1–1.6 μm optical communication systems," *Electronics Letters*, Vol. 16, 1980, pp. 750–751.

[31] Park, M. S., and R. A. Minasian, "Ultralow noise 10 Gb/s PIN/HEMT optical receiver," *IEEE Photonics Technology Letters*, Vol. 5, No. 2, 1993, pp. 161–162.

[32] Akatsu, Y., et al., "A 10 Gb/s high sensitivity, monolithically integrated PIN-HEMT optical receiver," *IEEE Photonics Technology Letters*, Vol. 5, No. 2, 1993, pp. 163–165.

[33] Muramoto, Y. et al., "High-speed monolithic receiver OEIC consisting of a waveguide PIN photodiode and HEMT's," *IEEE Photonics Technology Letters*, Vol. 7, No. 6, 1995, pp. 685–687.

[34] Lunardi, L. et al., "20 Gb/s monolithic PIN/HBT photoreceiver module for 1.55 μm applications," *IEEE Photonics Technology Letters*, Vol. 7, No. 10, 1995, pp. 1201–1203.

[35] Runge, P. K., "An experimental 50 Mb/s fiber optic PCM repeater," *IEEE Transactions on Communications*, Vol. COM-24, 1976, pp. 413–418.

[36] Musaka, W. M., "An experimental optical fiber link for low bit rate applications," *Bell System Tech Journal*, 1977, pp. 65–75.

[37] Kasper, B. L., J. C. Campbell, A. H. Gnauck, A. G. Dentai, J. E. Bowers, J. R. Talman, and W. S. Holden "An APD/FET optical receiver operating at 8 Gb/s," *IEEE Journal of Lightwave Technology*, Vol. LT-5, 1987, pp. 344–347.

[38] Smith, D. R., R. C. Hooper, P. P. Smith, and D. Wake, "Experimental comparison of a germanium APD and InGaAs PINFET receiver for longer wavelength optical communication systems," *Electronics Letters*, Vol. 18, 1982, pp. 453–454.

[39] Kasper, B. L., J. C. Campbell, A. H. Gnauck, A. G. Dentai, and J. R. Talman, "SAGM APD receiver for 2 Gb/s and 4 Gb/s, *Electronics Letters*, Vol. 21, 1985, pp. 982–984.

[40] Walker, S. D., and L. C. Blank, "Ge APD/GaAs FET op-amp transimpedance optical receiver design having minimum noise and intersymbol characteristics," *Electronics Letters*, Vol. 20, 1984, pp. 808–809.

[41] Honeywell Inc., *Lightwave*, Vol. 9, No. 4, 1992, p. 71.

[42] Hewlett-Packard, *Lightwave*, Vol. 9, No. 4, 1992, p. 70.

[43] AMP Inc., *Lightwave*, Vol. 9, No. 4, 1992, p. 67.

[44] Laser Diode Inc., *Lightwave*, Vol. 9, No. 4, 1992, p. 72.

[45] Tonguz, O. K. and L. G. Kazovsky, "Theory of direct detection lightwave receivers using optical amplifiers," *IEEE Journal of Lightwave Technology*, Vol. LT-9, 1991, pp. 174–181.

[46] Mukai, T., Y. Yamamoto, and T. Kimura, "S/N and error rate performance in AlGaAs semiconductor laser preamplifier and linear repeater systems," *IEEE J. Quantum Electron.*, Vol. QE-18, 1982, p. 1560.

[47] Henry, P. S., "Error rate performance of optical amplifiers," in *Tech. Dig. Opt. Fiber Commun. Conf. '89*, Houston, THK3.

[48] Olsson, N. A., "Lightwave systems with optical amplifiers," *IEEE Journal of Lightwaver Technology*, Vol. LT-7, 1989, pp. 1071–1082.

[49] Jacobs, I., " Effect of optical amplifier bandwidth on receiver sensitivity,." *IEEE Transactions on Communications*, Vol. 38, No. 10, 1990, pp. 1863–1864.

[50] Ramaswami, R., and P. A. Humblet, "Amplifier induced crosstalk in multichannel optical networks," *IEEE Journal of Lightwave Technology*, Vol. LT-8, 1990, pp. 1882–1886.

[51] Marcuse, D., "Derivation of analytical expressions for the bit-error probability in lightwave systems with optical amplifiers," *IEEE Journal of Lightwave Technology*, Vol. LT-8, 1990, pp. 1816–1823.

[52] Marcuse, D., "Calculation of bit-error probability for a lightwave system with optical amplifiers and post-detection Gaussian noise," *IEEE Journal of Lightwave Technology*, Vol. LT-9, 1991, pp. 505–513.

[53] Li, T., and M. C. Teich, "Bit-error rate for a lightwave communication system incorporating an erbium-doped fiber amplifier," *Electronics Letters*, Vol. 27, No. 7, 1991, pp. 598–600.

[54] Danielsen, S. L., "Detailed noise statistics for an optically preamplifier direct detection receiver," *IEEE Journal of Lightwave Technology*, Vol. 13, 1995, pp. 977–981.

[55] Edagawa, N. et al., "904 km, 1.2 Gb/s non-regenerative optical transmission experiment using 12 Er-doped fiber amplifiers," *ECOC'89*, (Gothenburg, Sweden), post deadline paper, pap. PD7-1.

[56] Edagawa, N., H. Taga, Y. Yoshida, M. Suzuki, S. Yamamoto, and H. Wakabayashi, "10 Gb/s, 1500 km transmission experiment using 22 Er-doped fiber amplifier repeaters," in *Tech. Dig., 17th EOOC*, Paris, France, 1991, paper PD20.

[57] Taga, H., N. Edagawa, H. Tanaka, M. Suzuki, S. Yamamoto, H. Wakabayashi, N. S. Bergano, C. R. Davidson, G. M. Homsery, D. J. Kalmus, P. R. Trischitta, D. A. Gray, and R. L. Maybach, "10 Gb/s, 9000 km IM-DD transmission experiments using 274 Er-doped fiber amplifier repeaters," in *Tech. Dig., OFC/IOOC'93*, San Jose, CA, 1992, Paper PD14.

[58] Wedding, B., and B. Franz, "Unregenerated optical transmission at 10 Gb/s via 204 km of standard single mode fiber using a directly modulated laser diode, *Electron Lett.*, Vol. 29, No. 4, 1993, pp. 402–404.

[59] Taga, H., N. Edagawa, Y. Yoshida, S. Yamamoto, M. Suzuki, and H. Wakabayashi, "10 Gb/s, 4500 km transmission experiment using 138 cascaded Er-doped fiber," in *Tech. Dig., OFC'92*, San Jose, CA, 1992, paper PD12.

[60] Taga, H., N. Edagawa, Y. Yoshida, S. Yamamoto, M. Suzuki, and H. Wakabayashi "The experimental study of the effect of fiber chromatic dispersion upon IM-DD ultra long distance optical communication systems using a 1000 km fiber loop," *IEEE Journal of Lightwave Technology*, Vol. 12, 1994, pp. 1455–1461.

[61] Gloge, D., et al., "High speed digital lightwave communication using LEDs and PIN photodiodes, *Bell Sys. Tech Journal*, Vol. 59, 1980, pp. 1365–1382.

[62] Ogawa, K, "Analysis of mode partition noise in laser transmission systems," *IEEE J. Quantum Electronics*, Vol. QE-18, 1982, pp. 849–855.

[63] G. Grosskopf et al., "Laser mode partition noise in optical wideband transmission links," *Electron Lett.*, Vol. 18, 1982, pp. 493–494.

[64] Wenke, G., and G. Elze, "Investigation of optical feedback effects on laser diodes in broadband optical transmission systems," *Journal of Optical Commun.*, Vol. 2, 1981, pp. 128–133.

[65] Cartledge, J. C., and A. F. Elrefaie, "Effect of chirping-induced waveform distortion on the performance of direct detection receivers using traveling-wave semiconductor amplifiers," *IEEE Journal of Lightwave Technology*, Vol. 9, 1991, pp. 209–219.

[66] O'Reilly, J. J. and H. J. Silva, "Chirp induced penalty in optical fiber systems," *Electronics Letters*, Vol. 23, No. 19, 1987, pp. 992–993.

[67] O'Reilly, J. J., J. R. F. da Rocha, and K. Schumacher, "Influence of timing errors on the performance of direct-detection optical fiber communication systems," *IEE Proceedings*, Vol. 132, Part J, No. 6, 1985, pp. 308–313.

Chapter 5

Coherent Detection Lightwave Systems

In the previous chapter, we studied the principles and applications of a direct detection lightwave receiver. Unlike a direct detection receiver, a coherent receiver combines the received signal with a signal that is locally generated and then processes the combined signal. The resulting photocurrent is a replica of the original signal, translated down in frequency from the optical domain to radio domain. It is important to note that *coherent detection* in electrical communication systems mainly refers to synchronous detection (the received signal is in phase with the local oscillator signal). In lightwave systems, coherent detection may involve either synchronous or asynchronous detection of the received signal. The coherent detection method offers significant improvements in receiver sensitivity and utilizes fiber bandwidth more efficiently than direct detection. The coherent lightwave systems were explored during the late 1980s and early 1990s at many research laboratories.

The subject of coherent detection lightwave systems has been discussed in recent books [1,2] and review articles [3,4]. In this chapter, we discuss the basic concepts as well as recent advances in coherent lightwave systems. Section 5.1 discusses the principle of operation of a coherent detection lightwave system. The typical coherent lightwave system configurations are discussed in Section 5.2, and Section 5.3 describes their performance degradation issues. The experiments and field trials are outlined in Section 5.4. Finally, Section 5.5 gives a comparative summary of the coherent lightwave systems with direct detection systems. The coherent WDM and SCM systems, including high-speed optical networks, are discussed in Chapter 8.

5.1 PRINCIPLES OF COHERENT LIGHTWAVE SYSTEMS

The basic principle of operation involved in a coherent lightwave system is to combine the received signal with the local oscillator signal, before it is detected by a photodetector. Figure 5.1 shows the basic principle of operation of a coherent lightwave receiver.

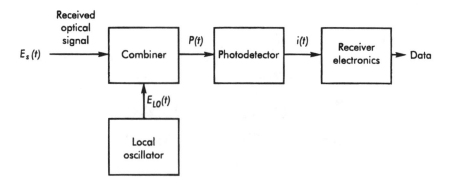

Figure 5.1 Basic principle of operation of a coherent lightwave receiver.

The electric fields associated with the received and local oscillator signals are represented by [1]

$$E_S(t) = \sqrt{2P_S} \cos(\omega_S t + \phi_S) \tag{5.1a}$$

$$E_{LO}(t) = \sqrt{2P_{LO}} \cos(\omega_{LO} t + \phi_{LO}) \tag{5.1b}$$

where $E_S(t)$ and $E_{LO}(t)$ represent the received and local oscillator signals respectively. The term ω_{LO} is the local oscillator frequency and ω_S is the signal frequency. The terms P_S and P_{LO} denote the received signal power and local oscillator power respectively. The terms ϕ_S and ϕ_{LO} denote the laser phase noise of the transmitter and receiver local oscillator respectively. We assume that the above electric fields are identically polarized. From (5.1), the combined signal power $p(t)$ is given by

$$P(t) = [\sqrt{2P_S} \cos(\omega_S t + \phi_S) + \sqrt{2P_{LO}} \cos(\omega_{LO} t + \phi_{LO})]^2 \tag{5.2}$$

Upon completing the square in (5.2) and by ignoring the second harmonics and the sum frequency components, the average optical power can be written as

$$P(t) = P_S + P_{LO} + 2\sqrt{P_S P_{LO}} \cos(\omega_{IF} t + \phi_S - \phi_{LO}) \tag{5.3}$$

where $\omega_{IF} = |\omega_S - \omega_{LO}|$ denotes the intermediate frequency as in radio systems. Now using $i(t) = \Re M P(t)$, the received signal current is given by

$$i(t) = \Re M(P_S + P_{LO}) + 2\Re M\sqrt{P_S P_{LO}} \cos(\omega_{IF} t + \phi_S - \phi_{LO}) + n_{LO}(t) \tag{5.4}$$

where $\Re$ denotes the photodetector responsivity and M is the avalanche multiplication factor of an APD. The term $n_{LO}(t)$ represents the additive shot noise of the local oscillator. In coherent lightwave systems, typically $P_{LO} >> P_S$ (strong local oscillator) is used, so that the shot noise due to the local oscillator is strong enough to overwhelm the other noise terms in the receiver. If $\omega_{IF} = 0$, then the local oscillator frequency exactly matches the signal frequency and the local oscillator phase is locked to the signal phase ($\phi_S = \phi_{LO}$), then the technique is known as the homodyne detection scheme. Assuming $P_{LO} >> P_S$ and ignoring the dc term, the resulting current for homodyne detection follows from (5.4) as

$$i(t) = 2\Re M\sqrt{P_S P_{LO}} + n_{LO}(t) \tag{5.5}$$

From (5.5), it can be seen that the average signal power can be increased by an order of magnitude by increasing the local oscillator power in comparison with the direct detection scheme, where $i(t) = \Re M P_S$. In practice, the design of homodyne receivers requires optical phase lock loop circuits, and it is difficult to achieve a good phase locking condition. However, the heterodyne detection receivers ($\omega_{IF} \neq 0$) are quite feasible, even though the sensitivity is degraded by 3 dB in contrast to the homodyne detection receivers.

Coherent lightwave systems also utilize optical amplifiers in long-haul systems, which eliminate the need for optoelectronic conversion [5–9]. For example, the use of optical amplifiers in WDM systems can offer significant performance advantages through improved capacity, reliability, and transparency. However, in an amplified system, the amplification of an optical signal is associated with the addition of amplifier spontaneous emission noise (ASE). With sufficient local oscillator (LO) power, the LO-ASE beat noise can be the dominant noise component in an amplified coherent lightwave system [8]. The main factor that degrades the performance of a coherent receiver is the phase noise of the semiconductor lasers. We will discuss the BER and sensitivity performance of coherent heterodyne receivers with and without the laser phase noise in the following subsections.

5.2 SYSTEM CONFIGURATIONS

This section reviews different configurations of coherent lightwave systems. The system configurations along with the practical receiver structures are discussed including their implementation aspects. In coherent lightwave systems, the information signal can be transmitted through amplitude, frequency, and phase modulation of the optical carrier wave, based on the following digital modulation schemes: (1) amplitude shift keying (ASK); (2) frequency shift keying (FSK); and (3) phase shift keying (PSK). The coherent lightwave systems are thus classified depending on the modulation and demodulation techniques used, and the various system configurations are shown in Figure 5.2. The ideal BER performance (shot noise limited) of these coherent lightwave receiver configurations is discussed in Section 5.2.1. Section 5.2.2 describes the performance of these receiver structures under the influence of laser phase noise. The coherent lightwave system using optical amplifiers is discussed in Section 5.2.3.

5.2.1 Receiver Structures

In this section, we describe some of the typical practical coherent lightwave receivers along with their ideal BER and sensitivity performance results. Specifically, we consider the coherent heterodyne receiver structures for asynchronous demodulation of ASK, FSK, and PSK modulated signals. The BER and sensitivity performance analysis of these receivers is carried out by assuming negligible laser phase noise. In the sensitivity analysis, we assume $P_{LO} >> P_S$, so that the LO shot noise $n_{LO}(t)$ in (5.3) is the only contributing noise source in the receiver. This noise source is assumed to be white Gaussian, whose single-sided power spectral density is given by [10]

$$N_{LO-Shot} = 2q\Re P_{LO} \tag{5.6}$$

where q is the electron charge.

5.2.1.1 ASK Systems

The implementation of an ASK transmission system requires the use of an external modulator, because the phase of the received signal should remain essentially constant for proper demodulation. This can be achieved by operating the semiconductor laser

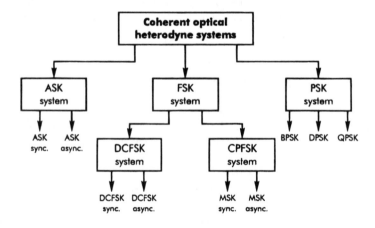

Figure 5.2 System configurations.

continuously at a constant current and modulating its output by external modulation. The external modulators such as Mach-Zehnder configuration [11] or semiconductor waveguide modulator [12] are utilized. The semiconductor waveguide modulator can be integrated with the semiconductor laser on the same chip and hence it is most suitable for the implementation of ASK transmission systems. The operation of these external modulators is described in Chapter 7.

The design of ASK receiving systems is based on either synchronous or asynchronous demodulation techniques. The synchronous demodulation is based on a correlation receiver structure, which requires very narrow linewidth lasers and the design of these receivers is quite complicated. In practice, the receiver structure based on asynchronous demodulation is often preferred, resulting in a much simpler receiver design. Figure 5.3 shows a block diagram of such a receiver using an envelope detection. When the laser transmitter is ASK modulated, the received signal current from (5.4) can be written by neglecting the dc term and the laser phase noise terms as [13]

$$i(t) = 2\Re M \sqrt{P_S P_{LO}} \cos(\omega_{IF}t) + n_{LO}(t) \; 0 \le t \ge T \tag{5.7a}$$

for bit 1 and

$$i(t) = n_{LO}(t) \; 0 \le t \ge T \tag{5.7b}$$

for bit 0 where T is the bit period. The local oscillator shot noise $n_{LO}(t)$ can be expressed in terms of in-phase and quadrature components as [13]

$$n_{LO}(t) = x(t)\cos \omega_{IF}t - y(t)\sin \omega_{IF}t \tag{5.8}$$

Now, the received signal current can be written by combining (5.7) and (5.8) as

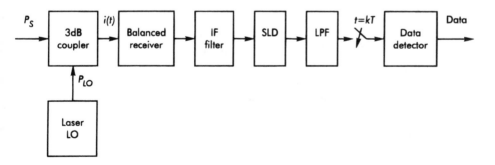

Figure 5.3 Block diagram of an ASK asynchronous (envelope) heterodyne receiver. (*Note:* SLD: square law detector; LPF: low-pass filter.)

$$i_T(t) = [2\Re M \sqrt{P_S P_{LO}} + x(t)]\cos \omega_{IF}t - y(t)\sin \omega_{IF}t \quad 0 \le t \ge T \quad (5.9a)$$

for bit 1 and

$$i_T(t) = x(t)\cos \omega_{IF}t - y(t)\sin \omega_{IF}t \quad 0 \le t \ge T \quad (5.9b)$$

for bit 0.

It is convenient to write the above equations in the envelope form as

$$i(t) = r(t, I_S)\cos[\omega_{IF}t + \theta(t)] \quad (5.10)$$

where

$$r(t, I_S) = \sqrt{[I_S + x(t)]^2 + y(t)^2} \quad (5.11a)$$

$$\theta(t) = \tan^{-1}\frac{y(t)}{I_S + x(t)} \quad (5.11b)$$

$$I_S = 2\Re M\sqrt{P_S P_{LO}} \quad (5.11c)$$

The probability density function (PDF) for (5.10) is well known in communication theory literature, which is called the Rice distribution. It is given by [13]

$$p(r, I_S) + \frac{r}{\sigma^2} I_0\left(\frac{rI_S}{\sigma^2}\right)\exp\left(-\frac{r^2 + I_S^2}{2\sigma^2}\right) \quad (5.12)$$

where I_S is equal to zero for the bit 0 and I_0 is a zeroth-order modified Bessel function of the first kind. Now the BER can be calculated as [13]

$$\text{BER} = \frac{1}{2}\left[1 - \int_{i_{TH}}^{\infty} p(r, I_S)dr\right] + \frac{1}{2}\int_{i_{TH}}^{\infty} p(r, 0)dr \quad (5.13)$$

The integral given in (5.13) can be calculated by using the Marcum's Q-function [13]

$$Q(\alpha_1, \beta_1) = \int_{\beta_1}^{\alpha_1} tI_0(\alpha_1 t)\exp\left(-\frac{t^2 + \alpha_1^2}{2}\right)dt \quad (5.14)$$

From (5.14), $Q(\alpha_1, 0) = 1$ and $Q(0, \beta_1) = \exp(-\beta_1/2)$. Hence, (5.13) can be written as

$$\text{BER} = \frac{1}{2}\left[1 - Q\left(\frac{I_S}{\sigma}, \frac{i_{TH}}{\sigma}\right)\right] + \frac{1}{2}\exp\left(-\frac{i_{TH}^2}{2\sigma^2}\right) \tag{5.15}$$

Now, the optimum threshold current can be found by differentiating the above expression with respect to i_{TH} and setting the result to zero yields

$$I_0\left(\frac{I_S i_{THO}}{\sigma^2}\right)\exp\left(-\frac{I_S^2}{2\sigma^2}\right) = 1 \tag{5.16}$$

where i_{THO} represents the optimum value of the threshold current which satisfies (5.16) and is given as

$$i_{THO} = \frac{I_S}{2} \tag{5.17}$$

Also, for large values of α_1 and β_1 (high SNR conditions), the Marcum Q-function can be approximated as [13]

$$Q(\alpha_1, \beta_1) = 1 - \frac{1}{2}\,\text{erfc}\left(\frac{\alpha_1 - \beta_1}{\sqrt{2}}\right) \tag{5.18}$$

Using (5.17) and (5.18) in (5.15), the expression form BER can be approximated as

$$\text{BER} = \frac{1}{4}\,\text{erfc}\left(\frac{I_S}{2\sqrt{2}\sigma}\right) + \frac{1}{2}\exp\left(-\frac{I_S^2}{8\sigma^2}\right) \tag{5.19}$$

Furthermore, for sufficiently large values of x, the erfc can be approximated as [13]

$$\text{erfc}(x) = \frac{\exp(-x^2)}{x\sqrt{\pi}} \tag{5.20}$$

Using (5.20) in (5.19) and noting that the first term is negligible for large values of argument, the BER can be written as

$$\text{BER} \approx \frac{1}{2}\exp\left(-\frac{I_S^2}{8\sigma^2}\right) \approx \frac{1}{2}\exp\left(-\frac{\gamma_{IF}}{4}\right) \tag{5.21}$$

where $\gamma_{IF} = I_S^2/2\sigma^2$ denotes the peak IF signal-to-noise-ratio of the system. From

(5.21), it can be inferred that the receiver sensitivity only degrades by 0.4 dB in comparison with the synchronous ASK receiver.

5.2.1.2 FSK Systems

The FSK transmission system is based on two approaches: (1) the direct or discontinuous-phase FSK (DCFSK) scheme and (2) the continuous-phase FSK (CPFSK) scheme. The DCFSK signal can be obtained by using two lasers with different frequencies for transmitting the binary bit stream via direct modulation. On the other hand, the CPFSK signal can be obtained via direct modulation, but with the use of only a single laser diode. The distributed feedback (DFB) or distributed brag reflector (DBR) semiconductor lasers are often used to achieve the direct modulation of the bit stream, because of their ability to operate in a single longitudinal mode with a narrow linewidth. Additionally, multielectrode DFB lasers can be utilized to obtain the uniform modulation response [14,15].

The demodulation of a DCFSK signal can be accomplished based on either synchronous or asynchronous demodulation techniques as similar to ASK systems. The synchronous demodulation generally uses a correlation receiver structure based on integrate and dump filters, whereas the asynchronous demodulation scheme utilizes dual band pass filters along with envelope detectors. Although a single filter receiver can be used for FSK, it results in a 3 dB sensitivity penalty. The block diagram of a DCFSK heterodyne receiver employing the dual filter and envelope detection is shown in Figure 5.4. In this case, the signals are represented by [13]

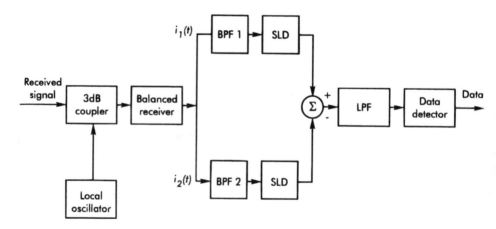

Figure 5.4 Block diagram of a dual filter DCFSK asynchronous heterodyne receiver.

$$i_1(t) = I_S \cos(\omega_1 t) + n_{LO}(t) \quad 0 \le t \le T \tag{5.22a}$$

for bit 1 and

$$i_2(t) = I_S \cos(\omega_2 t) + n_{LO}(t) \quad 0 \le t \le T \tag{5.22b}$$

for bit 0. The demodulation of the above signals is similar to that of ASK envelope detection, except that in this receiver structure, there are two envelope detector circuits, one on each branch. When a bit 1 is transmitted, the PDF of the demodulated signal r_1 in the first branch can be written as [13]

$$p_1(r_1) = \frac{r_1}{\sigma^2} I_0\left(\frac{r_1 I_S}{\sigma^2}\right) \exp\left(-\frac{r_1^2 + I_S^2}{2\sigma^2}\right) \tag{5.23a}$$

Similarly, when a bit 0 is transmitted, the PDF of the demodulated signal r_2 in the second branch is given by [13]

$$p_2(r_2) = \frac{r_2}{\sigma^2} \exp\left(-\frac{r_2^2}{2\sigma^2}\right) \tag{5.23b}$$

which is called the Raleigh distribution. Now, an error occurs when $r_1 < r_2$, so that the BER is calculated as [13]

$$\text{BER} = \int_{r_1=0}^{\infty} p_1(r_1)\left[\int_{r_2=r_1}^{\infty} p_2(r_2)\, dr_2\right] dr_1 \tag{5.24}$$

$$= \int_0^{\infty} \frac{r_1}{\sigma^2} I_0\left(\frac{r_1 I_S}{\sigma^2}\right) \exp\left(-\frac{r_1^2}{\sigma^2} - \frac{I_S^2}{2\sigma^2}\right) dr_1 \tag{5.25}$$

Normalizing equation (5.24) with respect to $t = \sqrt{2r_1}/\sigma$, we get

$$\text{BER} = \int_0^{\infty} \frac{t}{2} I_0\left(\frac{t I_S}{\sqrt{2}\sigma}\right) \exp\left(-\frac{t^2}{2} - \frac{I_S^2}{2\sigma^2}\right) dt \tag{5.26}$$

$$= \frac{1}{2} \exp\left(-\frac{\gamma_{IF}}{2}\right) \int_0^{\infty} t I_0(\sqrt{\gamma_{IF}}\, t) \exp\left(-\frac{t^2 + \gamma_{IF}}{2}\right) dt \tag{5.27}$$

Using (5.14), (5.27) can be written as

$$BER = \frac{1}{2} \exp\left(-\frac{\gamma_{IF}}{2}\right) Q(\sqrt{\gamma_{IF}}, 0) \tag{5.28}$$

Applying $Q(\sqrt{\gamma_{IF}}, 0) = 1$ to (5.28), we get

$$BER = \frac{1}{2} \exp\left(-\frac{\gamma_{IF}}{2}\right) \tag{5.29}$$

which is the expression for BER of dual filter DCFSK asynchronous system. It is important to note that the receiver sensitivity of a DCFSK single filter detection system degrades by 3 dB, resulting in (5.21) the same as that of ASK envelope detection systems.

Next, let us consider the demodulation of a CPFSK signal. The CPFSK signal with a modulation index of 0.5 (MSK scheme) can be demodulated by using a Costas PLL circuit [16], which requires stringent laser linewidths. The laser linewidths can be relaxed by using a differential detection scheme. The differential detection uses a delay-and-multiply receiver structure which operates as a frequency discriminator as shown in Figure 5.5 along with the frequency-to-voltage conversion characteristics.

Let us look at the principle of operation of a delay and multiply demodulator. The IF bandpass filter is assumed to be sufficiently wide so that no signal distortion occurs. The low pass filter eliminates all the second harmonics that are generated by the multiplier. The detected signal current at the noise limit in two channels of the demodulator can be expressed as [13]

$$i_1(t) = I_S \cos(\omega_{IF} t) \tag{5.30a}$$

$$i_2(t) = I_S \cos(\omega_{IF} t + \tau) \tag{5.30b}$$

where τ is the delay time of the demodulator. Now, the signal at the low pass filter output can be written as

$$i(t) = i_1(t) i_2(t) = b(t) \cos[\phi(t)] \tag{5.31}$$

where $b(t)$ is the amplitude and $\phi(t)$ is given by [17]

$$\phi(t) = \omega_{IF} \tau + b_k \frac{\Delta\omega}{2} \tau + \varphi(t) \tag{5.32}$$

where $\omega_{IF} = (2k + 1)\pi/2\tau$, b_k is the binary data taking on the value of $+1$ or -1

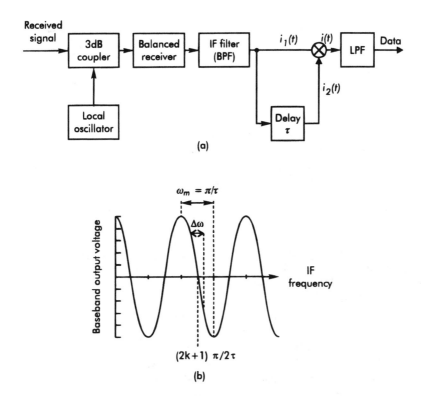

Figure 5.5 (a) Block diagram of a CPFSK differential detection receiver and (b) frequency-to-voltage conversion characteristics. Here, ω_m denotes the maximum angular frequency deviation of the signal. (*Source:* [17]. ©1987 IEEE. Reprinted with permission.)

with equal probability. The terms $\Delta\omega$ and $\varphi(t)$ represent the angular frequency deviation and the phase noise arising from the shot noise respectively. The data decision is performed based on the polarity of the output signal. The polarity of the output signal in turn depends on $\phi(t)$.

From the frequency-voltage characteristics, it can be seen that the output signal of the demodulator varies directly with the frequency of the input signal. The resulting frequency difference between the maximum and minimum values of $i(t)$ is $1/2\tau$. Hence the delay time of the demodulator can be determined based on the maximum and minimum frequency values of $i(t)$. The delay time is given by [17]

$$\tau = \frac{1}{2mB} \tag{5.33}$$

where B is the bit rate of the signal and m is the modulation index of the CPFSK signal. The resulting BER of the differential detection scheme is well known in communication theory literature and is given by [13]

$$BER = \frac{1}{2} \exp[-\gamma_{IF}] \tag{5.34}$$

where γ_{IF} is the shot noise–limited IF SNR. From (5.34), one can conclude that the differential phase shift keying (DPSK) system offers a 3 dB advantage in receiver sensitivity in comparison with a dual filter FSK system and a 6 dB advantage over an ASK envelope detection scheme.

5.2.1.3 PSK Systems

In the case of a binary phase shift keying (BPSK) system, the optical bit stream is generated by modulating the phase of the optical carrier wave, resulting in two values of phases (0 and 180 degrees) while the amplitude of the carrier wave remains constant. The implementation of BPSK requires an external modulator. Lithium Niobate based phase modulators can be used, but recently, multiple quantum well (MQW) phase modulators have been developed [18] for higher bit rate operation. The demodulation of the BPSK signal requires that the phase of the optical carrier remain stable so that the phase information can be extracted at the receiver without ambiguity. The synchronous demodulation receiver structure based on a Costas PLL circuit is often used [19]. However, this dictates a stringent condition on the laser linewidths.

The linewidth requirement can be relaxed to some extent by the use of a differential phase-shift keying (DPSK) technique similar to the CPFSK differential detection. In the case of DPSK, the bit stream is generated by using the phase difference between two neighboring bits. The advantage of DPSK over BPSK is that the phase of the carrier wave essentially remains constant over a duration of two bits. The DPSK receiver structure is the same as that of a differential CPFSK detection scheme as shown in Figure 5.5 and the expression for BER is given by (5.34).

Optical quadrature phase shift keying (QPSK) system can also be realized, whose configuration is similar to that for conventional BPSK optical heterodyne system [20]. The transmission system requires an external optical phase modulator and much stringent laser linewidths. The QPSK signal can be recovered at the receiver by using synchronous demodulation based on a Costas PLL circuit or by using an asynchronous demodulation using a delay-multiply receiver structure. Theoretically, the QPSK synchronous system offers the same receiver sensitivity as the synchronous BPSK system, but with an increase in bandwidth efficiency. Whereas the asynchronous QPSK system incurs a power penalty of 2.5 dB in comparison with the synchronous detection, it has a simpler configuration and can tolerate a larger laser phase noise.

Next, let us compare the ideal BER and sensitivity performance of different system configurations. It is often convenient to measure the sensitivity of receivers in terms of the number of photons/bit. The shot noise limited IF SNR can be found from the signal and noise components assuming a PIN diode ($M = 1$) as

$$\gamma_{IF} = \frac{I_S^2}{2\sigma^2} = \frac{4\mathcal{R}^2 P_S P_{LO}}{2(N_{LO-Shot}B_{IF})} = \frac{2\mathcal{R}^2 P_S P_{LO}}{2q\mathcal{R}P_{LO}B_{IF}} = \frac{\mathcal{R}P_S}{qB_{IF}} = \eta N_{P/bit} \qquad (5.35)$$

where $N_{p/bit} = P_S/h\nu B$ and B is the system data rate which is assumed the same as the IF bandwidth B_{IF}. In practice, it is important to note that the IF bandwidth of the receiver mainly depends on the spectrum of the modulated signal and the laser phase noise. If $\eta = 1$ in (5.35), then $\gamma_{IF} = N_{P/bit}$. Table 5.1 gives a comparative summary of the ideal BER and sensitivity results for different system configurations. The receiver sensitivity quoted in Table 5.1 refers to the average number of photons per bit to achieve a BER of 10^{-9}. The average number of photons per bit is reduced by a factor of 2 for coherent homodyne detection.

5.2.2 Impact of Laser Phase Noise

The main source of sensitivity degradation in coherent lightwave systems is the laser phase noise associated with the laser sources in the transmitter and receiver. The effect of laser phase noise is characterized by the IF linewidth-to-bit-rate ratio, where the IF linewidth is the sum of the transmitter and local oscillator linewidths. In general, the IF linewidth-to-bit-rate ratio depends on the modulation format and the demodulation technique. Generally, the asynchronous receivers have a higher degree of tolerance for the linewidth requirements than the synchronous receivers.

Table 5.1
Ideal BER and Sensitivity of Coherent Detection Lightwave System

System	BER	Sensitivity (N_P)	Sensitivity $(IF\ SNR\ dB)$
ASK synchronous	$0.5\ \text{erfc}(N_P/4)$	72	18.6
ASK asynchronous	$0.5\ \exp(-N_P/4)$	80	19.0
FSK synchronous	$0.5\ \text{erfc}(N_P/2)$	36	15.6
FSK asynchronous	$0.5\ \exp(-N_P/2)$	40	16.0
CPFSK asynchronous	$0.5\ \exp(-N_P)$	20	16.0
MSK synchronous	$0.5\ \text{erfc}(\sqrt{N_P})$	18	12.6
PSK synchronous	$0.5\ \text{erfc}(\sqrt{N_P})$	18	12.6
PSK asynchronous	$0.5\ \exp(-N_P)$	20	13.0
QPSK synchronous	$0.5\ \text{erfc}(\sqrt{N_P})$	18	12.6

The study related to the effect of laser phase on the BER performance of different coherent lightwave systems has been carried out by many workers using several methods [10,17,19,21–30]. The comparison of several methods is carried out in [21]. For example, in one study [22], numerical and computer simulation techniques were used to study the effect of laser phase noise on BER and sensitivity in asynchronous ASK and FSK heterodyne systems, employing the square law envelope detection and post detection filtering. The optimized detection performance in terms of the IF bandwidth and threshold voltage has been obtained. In this exact model, the impulse response of the IF filter is assumed as a finite time integrator and the post detection filter is treated as a discrete-time integrator. For the same model, Gaussian approximation technique was used to obtain a closed form expression for the BER in terms of IF SNR, laser linewidth, and IF bandwidth [10]. For this exact model, a rigorous analysis of the effects of laser phase noise was also carried out in [2].

In the following paragraphs, we consider an approximate model that consists of an ideal envelope detection for heterodyne asynchronous ASK and dual filter FSK systems [24,25]. We use Gaussian approximation for the laser phase noise distribution. Hence, the probability density function (PDF) of the laser phase noise fluctuations is modeled by [25]

$$p(\Delta\phi) = \frac{1}{\sqrt{2\pi\sigma_\phi^2}} \exp\left(-\frac{\Delta\phi^2}{2\sigma_\phi^2}\right) \qquad (5.36)$$

where $\Delta\phi$, represents the laser noise phase fluctuations and σ_ϕ^2 denotes the variance of $\Delta\phi$, which can be found in terms of the laser linewidth defined at the full width at half-maximum (FWHM) as [25]

$$\sigma_\phi^2 = 2\pi(\Delta\nu_{LT} + \Delta\nu_{LO})\tau_m = 2\pi\Delta\nu_{IF}\tau_m \qquad (5.37)$$

where $\Delta\nu_{LT}$ and $\Delta\nu_{LO}$ are the FWHM linewidths of the transmitter and local oscillator lasers, τ_m is the measurement interval, and $\Delta\nu_{IF}$ is the IF linewidth. Now defining the angular frequency fluctuation of the laser as a random process by $\Delta\omega = \Delta\phi/\tau_m$, we can write an expression for the PDF of the IF signal at the beat frequency as [25]

$$p_{IF}(\Delta\omega)d(\Delta\omega) = \frac{1}{\sqrt{4\pi^2\Delta\nu_{IF}/\tau_m}} \exp\left(-\frac{\Delta\omega^2\tau_m}{4\pi\Delta\nu_{IF}}\right)d(\Delta\omega) \qquad (5.38)$$

Note that the IF signal output is a random process having a PDF given by (5.38). Using the following normalization variables $\Delta\nu = \Delta\nu_{IF}/B$ and $\Delta u = \Delta\omega/2\pi B$, (the laser linewidths and the angular frequency variables are normalized with respect to the bit rate B), (5.38) can be written as

$$p_{IF}(\Delta u) = \frac{1}{\sqrt{\Delta v B_N}} \exp\left(-\frac{\Delta u^2 \pi}{\Delta v B_N}\right) \qquad (5.39)$$

where $p_{IF}(\Delta u)$ denotes the PDF for the deviation of the measured IF from the nominal IF (Δu), and $B_N = B_{IF}/B$ is the normalized IF bandwidth. Now, a general expression for the BER can be written taking into account the laser phase noise as [25]

$$\text{BER} = \int_{-\infty}^{\infty} P_e(P_S/\Delta u)p_{IF}(\Delta u)d(\Delta u) \qquad (5.40)$$

where $P_e(P_S/\Delta u)$ is the conditional error probability as a function of Δu and the received signal power. In the following sections, (5.40) forms the basis for evaluating the BER for different system configurations.

5.2.2.1 ASK Heterodyne Asynchronous System

When the ASK modulated signal is demodulated, the instantaneous frequency of the IF signal deviates from the nominal frequency due to the laser phase noise of the signal, requiring a higher IF bandwidth to accommodate the signal, and hence degrades the IF SNR. Hence, the amplitude of the demodulated signal after the envelope detector might decrease and become smaller than the threshold voltage, which causes errors. In the limit of large signal power (under infinite IF SNR conditions), the other receiver noise sources can be ignored and one can consider only the impact of laser phase noise through (5.40). In this limit, the conditional error probability P_e assumes a constant value of unity and the BER is determined by $p_{IF}(\Delta u)$. Hence, the asymptotic BER can be determined numerically by using [25]

$$\text{BER}_{\text{asym}} = \frac{1}{2} \int_{-\infty}^{\Delta u_-} p_{IF}(\Delta u)\,du + \frac{1}{2} \int_{\Delta u_+}^{\infty} p_{IF}(\Delta u)\,du \qquad (5.41)$$

$$= \frac{1}{4}\left[\text{erfc}\left(|\Delta u_-|\sqrt{\frac{\pi}{\Delta v B_N}}\right) + \text{erfc}\left(\Delta u_+\sqrt{\frac{\pi}{\Delta v B_N}}\right)\right] \qquad (5.42)$$

where Δu_- and Δu_+ are the negative and positive frequency deviations of the IF signal for which the output signal voltage for a received one is equal to the threshold voltage. From (5.42), it can be seen that the asymptotic error-rate mainly depends on the IF linewidth Δv and the IF bandwidth B_N and also implicitly on the threshold setting V_{TH} through Δu_- and Δu_+. The BER performance of an ASK heterodyne asynchronous system is shown in Figure 5.6. Curve a is the shot noise limit with zero linewidth lasers, with large local oscillator power and $B_N = B$.

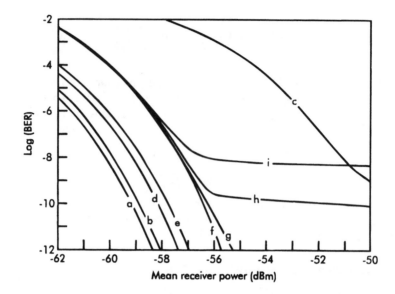

Figure 5.6 BER performance of an ASK heterodyne asynchronous system for various normalized linewidths and IF bandwidths. Curves *a–c* are for the ideal linear envelope detector with V_{TH} = 0.5. Curves *d–i* are for a square-law detector with post detection filtering and optimum threshold setting for $\Delta \nu = 0$. (*Source:* [24]. ©1987 IEEE. Reprinted with permission.)

The curve b is the same as curve a except that the local oscillator power is 0.1 mW. For curve d, the local oscillator power is large, $B_{IF} = 2B$ and curve e is the same as curve d but with 0.1 mW local oscillator power. The curves f to i are similar to curve e but with $B_{IF} = 5B$ and various linewidths (0–0.5). From Figure 5.6, it can be seen that the BER degrades as the linewidth increases. For $\Delta \nu = 0$–0.15 the incurred sensitivity penalty ranges from 3 to 6 dB [24]. However, for a square law detector with post detection filtering (exact model) system, the resulting sensitivity penalty varies from 0 to 2.2 dB over $\Delta \nu = 0$–2.5 [10,22].

5.2.2.2 FSK Heterodyne Dual Filter Detection System

In a dual filter FSK system, for large signal powers, an error occurs when the frequency deviation is so large that more signal passes through the wrong filter than through the correct filter, which happens if $\Delta u > m/2$ for a received zero or $\Delta u < m/2$ for a received one. In this case, the asymptotic BER can be evaluated for high IF SNR conditions as similar to the previous section as [25]

$$BER_{asymp} = \frac{1}{2}\int_{-\infty}^{-m/2} p_{IF}(\Delta u)\ du + \frac{1}{2}\int_{m/2}^{\infty} p_{IF}(\Delta u)\ du \qquad (5.43)$$

where m is the modulation index in FSK modulated system, which is defined as

$$m = \frac{(\omega_1 - \omega_0)}{2\pi B} \qquad (5.44)$$

where ω_1 and ω_0 are the signal frequencies for bit 1 and 0 respectively. Thus in the limit of infinite IF SNR, (5.43) reduces to

$$BER_{asymp} = \frac{1}{2}\operatorname{erfc}\left(\frac{m}{2}\sqrt{\frac{\pi}{\Delta\nu B_N}}\right) \qquad (5.45)$$

Here, the BER depends on the modulation index, normalized IF linewidth and IF bandwidth. The typical BER characteristics of a dual filter FSK asynchronous system are shown in Figure 5.7. It can be seen that the BER degrades as the linewidth increases as in the case of ASK heterodyne asynchronous system. For $\Delta\nu = 0.1–0.3$,

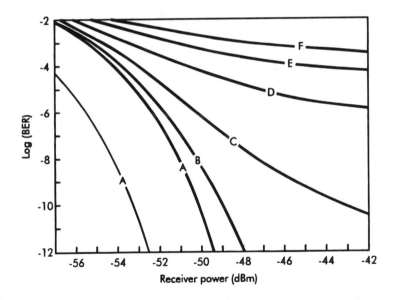

Figure 5.7 BER characteristics of a dual filter 400 Mb/s FSK heterodyne asynchronous system. Here, the thick curves A–F correspond to IF linewidths of 0, 0.1, 0.25, 0.5, 0.75, and 1 and the thin curve A refers to the strong local oscillator limit. (*Source:* [24]. ©1987 IEEE. Reprinted with permission.)

the incurred sensitivity penalty ranges from 2 to 8 dB [24]. However, for a square law detector with post detection filtering (exact model) system, the resulting sensitivity penalty varies from 0 to 3.0 dB over $\Delta\nu = 0$–2.5 [10,22].

5.2.2.3 CPFSK Heterodyne Asynchronous System

In CPFSK differential detection (delay and multiply receiver) system, the BER can be found by using [17]

$$\text{BER} = \int_{-(\Delta\omega/2)\tau}^{\pi-(\Delta\omega/2)\tau} \int_{-\infty}^{\infty} p_n(\phi_1 - \phi_2) p_{IF}(\phi_1) \, d\phi_1 \, d\phi_2 \tag{5.46}$$

where $p_n(\phi_1 - \phi_2)$ is the PDF of phase noise due to the shot noise, and $p_{IF}(\phi_1)$ is the PDF of the IF signal due to both the transmitter and local oscillator lasers, and τ is time delay of the differential detection. Equation (5.46) can be expressed as [17]

$$\text{BER} = \frac{1}{2} - \frac{\gamma_{IF} e^{-\gamma_{IF}}}{2} \sum_{n=0}^{\infty} \frac{(-1)^n}{(2n+1)} \left[I_n\left(\frac{\gamma_{IF}}{2}\right) \right. \tag{5.47}$$

$$\left. + I_{n+1}\left(\frac{\gamma_{IF}}{2}\right) \right]^2 e^{[-(2n+1)^2\pi\Delta\nu\tau]} \cos[(2n+1)\alpha_m]$$

where γ_{IF} is the shot noise limited SNR and $I_n(x)$ is the modified Bessel function of the first kind, and the term α_m is given by

$$\alpha_m = \frac{\pi(1 - \beta_m)}{2} \tag{5.48}$$

and β_m is the modulation index parameter defined by

$$\beta_m = \frac{\Delta\omega}{\omega_m} = \frac{2m\tau}{T_0} \tag{5.49}$$

where ω_m is the maximum angular frequency deviation of the signal, τ is the delay, m is the modulation index, and T_0 is the pulse width of the signal. Figure 5.8 shows the BER performance of the differential minimum shift keying (MSK) (CPFSK with $m = 0.5$) system for various linewidths. For a 1 dB power penalty, the required laser linewidth is approximately 3.3×10^{-3}. However, the required laser linewidth increases as the modulation index increases. The relationship between the required linewidth and the modulation index parameter is given by [17]

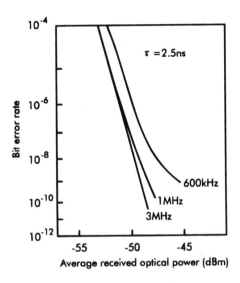

Figure 5.8 BER performance of a 400 Mb/s MSK versus average received signal power as a function of line widths. (*Source:* [17]. ©1987 IEEE. Reprinted with permission.)

$$\Delta \nu \tau < 10^{1.53\beta - 4} \qquad (5.50)$$

For $\tau = T_0/2m$, the required linewidth from (5.50) simplifies to $\Delta \nu T_0 < 6.8 \times 10^{-3} m$.

5.2.2.4 DPSK Heterodyne Differential Detection System

The BER characteristics of a DPSK heterodyne differential detection scheme is similar to that of a CPFSK differential detection scheme, which can be calculated from [17]

$$BER = 2 \int_{\pi/2}^{\pi} p_n(\phi_1 - \phi_2) p_{IF}(\phi_1) \, d\phi_1 \, d\phi_2 \qquad (5.51a)$$

$$BER = \frac{1}{2} - \frac{\gamma_{IF} e^{-\gamma_{IF}}}{2} \sum_{n=0}^{\infty} \frac{(-1)^n}{(2n+1)} \left[I_n\left(\frac{\gamma_{IF}}{2}\right) + I_{n+1}\left(\frac{\gamma_{IF}}{2}\right) \right]^2 e^{[-(2n+1)^2 \pi \Delta \nu T_0]} \qquad (5.51b)$$

From (5.51b), it can be seen that the BER characteristics of a DPSK heterodyne asynchronous system is the same as that of CPFSK asynchronous system with $m = 0.5$ and $\beta = 1$. Hence, the requirement for the IF linewidth is the same as that for an MSK differential asynchronous system.

The normalized IF linewidth requirements required for each system are summarized in Table 5.2 for a BER = 10^{-9} to incur a sensitivity penalty of less than 1 dB.

As seen from Table 5.2, the ASK and FSK asynchronous heterodyne systems can tolerate higher laser linewidths than the other systems. For example, in 2.5 Gbps systems, the IF linewidth requirement for an FSK heterodyne dual-filter detection scheme is approximately 500 MHz, and for a QPSK asynchronous system, the IF linewidth requirement is approximately 187.5 kHz. It is important to know that for lower data rates, the IF linewidth requirement becomes more stringent.

The laser phase noise problem can also be solved by using the phase diversity techniques for coherent lightwave systems. The phase diversity receiver is based on multiport techniques that use two or more photodetectors whose outputs are combined to yield a signal, which is independent of the phase difference, thus avoiding the phase locking condition. These phase diversity receivers offer robust and good performance for high-speed systems, but require complex circuits [31,32].

Table 5.2
Normalized IF Linewidth Requirements

System	Synchronous	Asynchronous	References
ASK	—	0.5 (exact model)	[10,22]
FSK dual filer	—	0.2 (exact model)	[10,22]
MSK	3.0×10^{-4}	3.3×10^{-3}	[17,29]
BPSK	2.26×10^{-3}	3.3×10^{-3}	[17,19]
QPSK	7.5×10^{-5}	2.5×10^{-4}	[30]

5.2.3 Coherent Detection with Optical Amplifiers

As stated in Chapter 3, the optical amplifiers can be utilized in three different system applications: (a) the booster system; (b) the preamplifier system; and (c) the in-line amplifier system. We will discuss these applications in coherent lightwave systems in the following paragraphs.

5.2.3.1 Booster System

The optical amplifier can be utilized to enhance or boost the signal output of a laser transmitter for extending the unrepeated transmission length. For example, the use of an EDFA with a gain of 20 dB, can extend the transmission length of an unrepeatered link as much as 100 km assuming a transmitter power output of 0 dBm and a single mode fiber. Additionally, the booster application of optical amplifiers can minimize the number of repeaters in a system design. In practice, however, the maximum input power to the optical fiber is limited by the effect of stimulated

Brillouin scattering (SBS). With a 20 dBm launched signal power, repeaterless transmission distance of 364.3 km has been achieved at 2.5 Gbps for CPFSK system [33]. Another application of an optical amplifier as a booster amplifier is to enhance the signal power of a local oscillator to achieve the shot-noise limited operation [9].

5.2.3.2 Preamplifier System

In a preamplifier system, the optical amplifier is used at the front end of a receiver. Ideally a coherent receiver should operate in shot-noise limited condition and the use of an optical preamplifier in coherent systems is ineffective. However, in practice, the realization of a shot-noise limited coherent receiver is a difficult task particularly at multi-gigabits per second data rates, because it is difficult to realize a low noise optical receiver with a wide bandwidth and in such a case the preamplifier application might be useful. The effectiveness of the optical amplifier as a preamplifier is demonstrated in Figure 5.9.

From Figure 5.9, it can be seen that at a loss of 0 dB, the SNR of the receiver with an optical preamplifier is degraded by 3 dB in comparison with the system without the amplifier, which implies that the shot-noise limited SNR is better than the LO-ASE beat noise limited SNR by 3 dB. From these results, we can see that the use of the preamplifier in a coherent lightwave system that operates in the shot-noise limited condition is ineffective. Also, from this figure, it can be inferred that if we use the optical amplifier as an in-line amplifier to compensate the loss before the receiver, we can extend the transmission distance of a coherent lightwave system. In the next section, we will discuss a model of a cascaded in-line amplifier system and study its effectiveness in coherent lightwave systems.

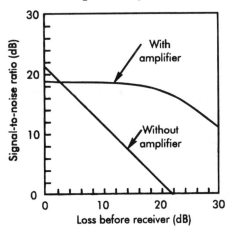

Figure 5.9 Signal-to-noise-ratio versus additional loss before the receive with and without an optical amplifier. (*Source:* [1]. Reprinted with permission.)

5.2.3.3 In-line Amplifier System

Consider a cascaded in-line amplifier system as shown in Figure 5.10. In this model, we assume n stage identical in-line optical amplifiers with the same gain and equal spacing. The loss of the optical fiber at each span is compensated by the gain of each amplifier. In this model, the effects of laser phase noise and fiber nonlinearites have not been considered. Additionally, the saturation effect of the amplifier is ignored. In this model, an expression for BER can be expressed in terms of the IF signal-to-noise ratio as [7]

$$\text{BER} = \frac{1}{2}\exp(-K\gamma_{IF}) \tag{5.52}$$

where K is a constant, which depends on the modulation/demodulation schemes. It is given by

$$K = \frac{1}{4} \tag{5.53a}$$

for the ASK asynchronous system,

$$K = \frac{1}{2} \tag{5.53b}$$

for the FSK dual filter system, and

$$K = 1 \tag{5.53c}$$

for the CPFSK/DPSK delay system, and the IF SNR is given by

$$\gamma_{IF} = \frac{S_P}{\sigma_T^2} \tag{5.54}$$

where $S_P = 2\Re^2 P_s P_{LO}$ is the signal power and the σ_T^2 is the total variance of the noise, which is given by [7]

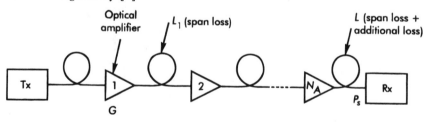

Figure 5.10 Model of a cascaded in-line amplifier system. (*Source:* [7]. © 1991 IEEE. Reprinted with permission.)

$$\sigma_T^2 = (N_{\text{LO-Shot}} + N_{\text{LO-SP}} + N_{\text{Sig-Shot}} + N_{\text{S-SP}} + N_{\text{SP}} + N_{\text{SP-SP}} + N_{\text{TH}})B_{IF} \quad (5.55)$$

where

$N_{\text{LO-Shot}}$ is the shot noise due to local oscillator light;

$N_{\text{LO-SP}}$ is the beat noise between the local oscillator light and the spontaneous emission;

$N_{\text{Sig-Shot}}$ is the shot noise due to the signal light;

$N_{\text{S-SP}}$ is the beat noise between the signal light and the spontaneous emission;

N_{SP} is the shot noise due to spontaneous emission;

$N_{\text{SP-SP}}$ is the beat noise of the spontaneous emission;

N_{TH} is the receiver circuit thermal noise.

The single sided power spectral density of the LO-shot noise term is given by (5.6). The single sided power spectral densities of the remaining terms are given by [7]

$$N_{\text{LO-SP}} = 4q\mathcal{R}P_{\text{LO}}F_{\text{SP}} \quad (5.56a)$$

$$N_{\text{Sig-shot}} = 2q\mathcal{R}P_S G \quad (5.56b)$$

$$N_{\text{S-SP}} = 4q\mathcal{R}P_S G F_{\text{SP}} \quad (5.56c)$$

$$N_{\text{SP}} = 4q^2 F_{\text{SP}}\Delta f \quad (5.56d)$$

$$N_{\text{SP-SP}} = 2q^2 F_{\text{SP}}^2 \Delta f \quad (5.56e)$$

$$N_{\text{TH}} = \frac{4k_B T_k (NF)}{R_L} \quad (5.56f)$$

where NF denotes the noise figure of an amplifier, Δf represents the bandwidth of an optical amplifier, k_B is the Boltzmann's constant, T_k is the temperature in Kelvin, and the noise factor F_{SP} is denoted by [7]

$$F_{\text{SP}} = \eta L L_1 n_{\text{SP}} N_A (G - 1) \quad (5.57)$$

where η denotes the quantum efficiency of the photodetector, N_A is the number of amplifiers, n_{SP} is the spontaneous emission factor of the amplifier, L is the loss between the last in-line amplifier and the receiver without L_1, and $L_1 = 1/G$ represents the span loss between the two successive in-line amplifiers. Using (5.52)–(5.57), the BER characteristics of the in-line amplifier system can be obtained. Using this model, one can estimate the power penalty of the system as a function of system length. Here, the power penalty is measured in terms of the receiver sensitivity degradation in dB for a BER $= 10^{-9}$, compared with the sensitivity without the amplifier. Figure 5.11 shows the power penalty as a function of system length at a data rate of 2.4 Gbps, including the intensity modulation direct detection (IM/DD) system.

It can be seen that the power penalty is quite significant in the case of a IM/DD system, whereas in coherent systems, about 1 dB power penalty can be achieved by using CPFSK or DPSK formats. This is due to the achievement of LO-ASE beat noise limited condition in coherent lightwave systems. Another advantage of using inline amplifiers in coherent systems is that a wide dynamic range can be obtained in comparison with the intensity modulation direct detection (IM/DD) system. At a 1 dB power penalty, the dynamic range of an ASK system is about 9 dB wider than that of IM/DD system using an APD with an avalanche multiplication factor of 40. In the case of CPFSK or DPSK systems, the improvement in system dynamic range is about 15 dB [7].

In the case of cascaded amplified systems, the spontaneous-emission noise for each amplifier propagates through the rest of the optical fiber and is amplified by successive amplifiers along with the signal. Thus, ASE can accumulate before reaching

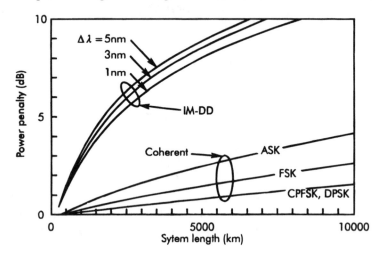

Figure 5.11 Power penalty versus the system length. Here, $\Delta\lambda$ denotes the bandwidth of an optical amplifier. In the calculation, input power of each amplifier was assumed to be -20 dBm, $G = 20$ dB, $\Delta f = 50$ nm, and $NF = 5$ dB. (*Source:* [7]. ©1991 IEEE. Reprinted with permission.)

the receiver and as the ASE level increases, it begins to saturate the optical amplifiers and decreases the signal gain and hence SNR in the system. The maximum number of in-line amplifiers is limited by the accumulated amplifier noise.

5.3 PERFORMANCE DEGRADATION ISSUES

The performance degradation issues, such as relative intensity noise (RIN), polarization mismatch, fiber chromatic dispersion, and nonlinear effects, play an important role in the design of practical coherent lightwave systems. These issues are discussed along with the system performance in the following subsections.

5.3.1 Polarization Mismatch

In coherent lightwave systems, it is important that the received signal and locally generated signal have identical polarizations. To obtain polarization match, several schemes including the polarization maintaining fiber, or adaptive polarization scheme, or even a more complicated polarization diversity receiver structure, can be utilized. The polarization mismatch problem has been extensively studied by several research workers [34–38]. In practice, the polarization diversity receivers or polarization control schemes can be utilized, because the cost of the polarization maintaining fiber is too high for long-haul systems. Also, the technique of polarization diversity can be combined with phase diversity to solve both phase and polarization fluctuations of the received signal [38].

5.3.2 Chromatic Dispersion

The chromatic dispersion affects the coherent lightwave systems in terms of the system bandwidth and the repeater spacing. However, the dispersion can be compensated, in principle, by using a dispersion compensating fiber, in addition to the use of delay equalizers in the IF domain. Also, the low loss dispersion shifted fibers in the 1.55 μm window can be used. The effect of fiber dispersion on the system performance has been studied by several workers [39–43] for various modulation formats. Figure 5.12 shows the dispersion penalty versus the chromatic dispersion index for various modulation formats [40]. In Figure 5.12, the system performance is characterized by the power penalty and the dispersion index. The chromatic dispersion index parameter γ is characterized by [40].

$$\gamma = \frac{1}{\pi} B^2 L D(\lambda) \frac{\lambda^2}{c} \qquad (5.58)$$

where B is the data rate, L is the fiber length, $D(\lambda)$ is chromatic dispersion parameter, c is the velocity of light, and λ is the optical frequency. In Figure 5.12, note that the

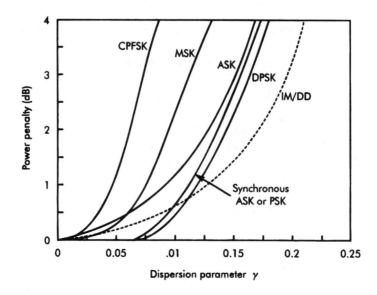

Figure 5.12 Dispersion penalty versus chromatic dispersion index for CPFSK, MSK, DPSK PSK, ASK, and IM/DD systems. (*Source:* [40]. ©1988 IEEE. Reprinted with permission.)

slope of the dispersion penalty curves is more steep for coherent systems than for OOK systems. When the value of λ is above 0.2, the penalty of OOK system is less than any of the coherent systems, because, the optical dispersion penalty for OOK systems is half the electrical power penalty. If γ is less than 0.2, PSK and DPSK systems introduce less penalty than OOK.

The results for a coherent ASK synchronous system are identical to that of the PSK case, because the baseband components for these systems are the same except for the dc term. Also, it can be seen that the power penalty can be reduced to less than 1 dB in most systems if the dispersion index is less than 0.1. Additionally, the direct detection OOK system allows the maximum transmission distance for a given modulation rate and chromatic dispersion, while the CPFSK system is the most severely affected.

5.3.3 Other Factors

The relative intensity noise problem can be solved by using the balanced receivers in coherent lightwave systems [44–46]. A balanced receiver employs dual photodetectors and a mixing circuit. The use of a balanced receiver uses the local oscillator power more efficiently, so that the receiver can be operated in the shot noise limit. Figure 5.13 shows the diagram of a balanced coherent receiver.

It consists of one beam splitter, two photodetectors, and one differential combiner. The signal and local oscillator signals are fed into each input port of the beam

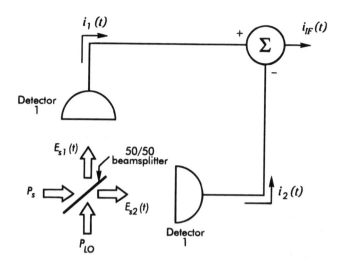

Figure 5.13 Balanced photodetector configuration for a coherent receiver. (*Source:* [44]. ©1987 IEEE. Reprinted with permission.)

splitter. Assuming a perfect splitting ratio of 50% for the beam splitter and equal electrical and optical lengths between the output of the beam splitter and the input of the differential combiner at each branch, the electric field of the signal at each branch can be written as

$$E_{S1}(t) = \sqrt{P_S(t)} \cos(\omega_S t + \phi_S - \pi/2) \tag{5.59}$$

$$E_{S2}(t) = \sqrt{P_S(t)} \cos(\omega_S t + \phi_S) \tag{5.60}$$

where $\pi/2$ in (5.69) represents the phase shift associated with the electric field of the signal in branch 1 relative to the signal field that passed through the beam splitter into branch 2. Using the similar situation, the local oscillator fields in each branch can be written as [44]

$$E_{LO1}(t) = \sqrt{P_{LO}(t)} \cos(\omega_{LO} t + \phi_S) \tag{5.61}$$

$$E_{LO2}(t) = \sqrt{P_{LO}(t)} \cos(\omega_{LO} t + \phi_{LO} - \pi/2) \tag{5.62}$$

From (5.59)–(5.62), the generated current at the output of each photodetector can be written as

$$i_i(t) = \frac{\mathcal{R}M}{2}(P_S(t) + P_{LO}(t) + 2\sqrt{P_S(t)P_{LO}(t)} \, [\sin(\omega_S - \omega_{LO})t$$

$$+ \, \phi_S - \phi_{LO}]) + n_{LO1}(t) \tag{5.63}$$

$$i_2(t) = \frac{\Re M}{2}(P_S(t) + P_{LO}(t) + 2\sqrt{P_S(t)P_{LO}(t)}\ [\sin(\omega_S - \omega_{LO})t$$

$$+ \ \phi_S - \phi_{LO}]) + n_{LO2}(t) \qquad (5.64)$$

where $\Re$ is the responsivity of each photodetector which is assumed to be the same for each branch. The terms $n_{LO1}(t)$ and $n_{LO2}(t)$ represent the shot noise in each branch the circuit. From (5.63) and (5.64), the resulting current at the output of a differential combiner be written as

$$i_{IF}(t) = 2\Re M\sqrt{P_S(t)P_{LO}(t)}\ [\sin(\omega_S - \omega_{LO})t + \phi_S - \phi_{LO}]$$

$$+ \ (n_{LO1}(t) - n_{LO2}(t)) \qquad (5.65)$$

From (5.65) it can be seen that the intensity fluctuations of the received signal and local oscillator signal can be effectively eliminated due to the differential operation of the balanced receiver. Thus by using a balanced photodetector configuration for a receiver, an improvement in receiver sensitivity of about 6 to 7 dB can be achieved in comparison with a single photodetector receiver configuration, because of the power loss due to the single branch [44]. However, in a balanced receiver, in practice, it is often difficult to match the characteristics of two photodetectors and also not easy to make the equal lengths for the two paths. In this case, the suppression capability of the common mode noise is limited and this can be characterized by the common mode rejection ratio (CMRR). The CMRR is defined as the improvement of the RIN due to the local oscillator light. A CMRR of more than 30 dB is usually required to achieve a nearly shot noise limited operation of the receiver.

Other degradation mechanisms include the reflection problems [47–49] and nonlinear effects [50–52]. The optical isolators are commonly used in systems to avoid the optical reflection problems. Nonlinear effects such as stimulated Brillouin scattering (SBS) can degrade the system performance if the launched power is above the threshold value [50]. The SBS threshold depends on the modulation format as well on the bit rate.

Another limiting factor in coherent lightwave systems for achieving the maximum transmission distance is the self-phase modulation due to Kerr effect in optical fibers. The Kerr effect is refractive index change proportional to the square of the electric field in optical fibers, which gives rise to self-phase modulation. In coherent FSK and PSK lightwave systems, the signal envelope is essentially constant, which yields a constant phase rotation along the whole time period of the data stream and this effect is less severe in IM/DD ASK system. In [51], it has been reported that the AM noise of the transmitted signal induced by spontaneous emission noise from optical amplifiers is converted into PM noise due to the nonlinear Kerr effect of optical fibers, that gives rise to the increase of the phase noise of the signal light, which in turn degrades the performance of coherent lightwave systems. The nonlinear Kerr effect can be minimized by the use of optical phase conjugation technique. In this technique, both the dispersion and nonlinearity at each position along the fibers are compensated by an optical phase conjugator, which can be located in the transmitting or the receiving terminal [52].

5.4 EXPERIMENTS AND FIELD TRIALS

During the late 1980s and 1990s, a very large number of laboratory experiments and field trials were conducted to study the feasibility of coherent lightwave systems. [33,53–62]. In this section, some of the recently published experimental results pertaining to 1.5 μm coherent lightwave systems with and without optical amplifiers are presented. The performance of the system in these experiments is evaluated in terms of the receiver sensitivity and the transmission distance. Some of these results are summarized in Table 5.3.

The transmission experiments for ASK, FSK, and DPSK asynchronous systems were also conducted at 4 Gbps [59] using tunable DFB and DBR semiconductor lasers. The transmission distance of 175 km was achieved using a dispersion shifted fiber. Also, the performance of 10-Gbps ASK, FSK, and DPSK coherent systems has been explored in [60,61]. The CPFSK system operating at 8 Gbps was reported to yield a transmission distance of 203 km. [62]. The 8-Gbps PSK experiment in [62] used a synchronous detection scheme, external cavity semiconductor lasers, and a microstrip-line delay equalizer. The current research is focused on coherent heterodyne systems operating at 10 Gbps and beyond.

5.5 COMPARISON OF COHERENT LIGHTWAVE SYSTEMS

In the past few years, the coherent heterodyne lightwave systems have proven to be attractive for long-haul systems and network applications [63]. On the other hand,

Table 5.3
List of Experiments and Field Trials

Experiment	Bit rate	Lab	Sensitivity	Distance (km)	References
MSK	560 Mbps	NTT	−35.5 dBm	90	w/o amp[53]
DPSK	565 Mbps	BTRL	−47.6 dBm	176	w/o amp[33]
CPFSK	2.488 Gbps	NTT	−39.0 dBm	431	w/o amp[54]
CPFSK	1.187 Gbps	NTT	−41.8 dBm	195	1 EDFA[55]
FSK	2.0 Gbps	BellCore	−36.7 dBm	101	w/o amp[56]
DPSK	2.0 Gbps	AT&T	−39.0 dBm	170	w/o amp[57]
CPFSK	2.5 Gbps	NTT	−42.0 dBm	2223	25 EDFA[58]

an ideal amplified direct detection scheme offers identical performance to that of an ideal coherent heterodyne detection scheme in addition to the low cost and simplicity [64,65]. It has been shown in [64] that the performance of a heterodyne receiver and a preamplified direct detection receiver is identical under similar operating conditions. This equivalency is independent of the modulation format used. However, the improvement of receiver technology based on optoelectronic integrated circuits may make coherent lightwave systems feasible as well as attractive in terms of the cost and simplicity for multigigabit WDM and SCM systems. Such systems are expected to be developed for telecommunication trunks, CATV networks, and broadband subscriber networks. Research is under way at many laboratories for developing multigigabit (above 10 Gbps) coherent heterodyne lightwave systems to meet the future demands of telecommunication traffic [66].

Problems

Problem 5.1

Derive an expression for SNR of a coherent homodyne receiver. Assume shot noise limited operation.

Problem 5.2

Calculate the sensitivity of a ASK homodyne receiver operating at 1.3 μm assuming the shot noise limited operation and a BER = 10^{-9}. Assume a PIN diode with an efficiency of 0.8 and bit rate of the signal as 1 Gbps.

Problem 5.3

Repeat Problem 5.2 for a PSK homodyne receiver.

Problem 5.4

Using the communication theory principles, derive an expression for the shot noise limited BER of the following lightwave receivers.
   a. ASK synchronous receiver;
   b. FSK synchronous receiver;
   c. BPSK synchronous receiver.

Problem 5.5

Calculate the sensitivity of a 1.55 μm ASK heterodyne asynchronous receiver assuming shot noise limited operation and a BER = 10^{-9}. Assume a bit rate of 2 Gbps and a PIN diode with an efficiency of 0.8.

Problem 5.6

Repeat Problem 5.5 for dual filter FSK, BPSK, and CPFSK asynchronous receivers.

Problem 5.7

Calculate the BER of a 140-Mbps ASK asynchronous receiver with an IF linewidth of 0.3. The negative and positive deviations of the IF signal are −0.5 and 0.5 respectively. Assume an IF bandwidth of 280 MHz.

Problem 5.8

Calculate the BER of a 200-Mbps FSK dual filter asynchronous receiver with an IF linewidth of 0.3. Assume an IF bandwidth of 0.5 GHz and a modulation index of 0.7.

Problem 5.9

Calculate the BER of a 400-Mbps CPFSK differential detection asynchronous receiver with an IF linewidth of 0.3. Assume a modulation index parameter of 0.7, with a modulation index of 0.5.

Problem 5.10

Repeat Problem 5.9 for a DPSK receiver.

Problem 5.11

Consider an in-line amplifier CPFSK 140-Mbps system with five stages of cascaded amplifiers, each having a gain of 20 dB, P_{in} = −35 dBm, and P_S = −50 dBm. Calculate the BER of the system given the following system parameters: λ = 1.55 μm; η = 0.8; P_{LO} = 10 dBm; n_{SP} = 1.4; $\Delta\lambda$ = 50 nm; L = 1.0 dB; T_k = 300K; and R_L = 10 kOhm.

Problem 5.12

Repeat Problem 5.11 for dual filter FSK detection scheme.

Problem 5.13

Repeat Problem 5.11 for ASK envelope detection scheme.

Problem 5.14

Derive an expression for a balanced PIN detector configuration of homodyne and heterodyne receivers.

References

[1] Ryu, S., *Coherent Lightwave Communication Systems*, Norwood: Artech House Publishers, 1994.

[2] Jacobsen, G., *Noise in Digital Optical Transmission Systems*, Norwood: Artech House Publishers, 1994.

[3] Salz, J., "Coherent Lightwave Communications," *AT&T Tech. Journal*, vol. 64, no. 10, 1985, pp. 2153–2209.

[4] "Special issue on coherent communications," *IEEE Journal of Lightwave Technology*, vol. LT-8, no. 3, 1990.

[5] Olsson, N. A., "Lightwave systems with optical amplifiers," *IEEE Journal of Lightwave Technology*, vol. 7, no. 7, 1989, pp. 1071–1082.

[6] Saito, S., T. Imai, and T. Ito, "An over 2200 km coherent transmission experiment at 2.5 Gb/s using erbium-doped fiber in-line amplifiers," *IEEE Journal of Lightwave Technology*, vol. 9, no. 2, 1991, pp. 161–169.

[7] Ryu, S., S. Yamamoto, H. Taga, N. Edagawa, Y. Yoshida, and H. Wakabayashi, "Long haul coherent optical fiber communication systems using optical amplifiers," *IEEE Journal of Lightwave Technology*, vol. LT-9, no. 2, 1991, pp. 251–260.

[8] Walker, G. R., N. G. Walker, R. C. Steele, M. J. Creamer, and M. C. Brain, "Erbium-Doped fiber amplifier cascade for multichannel coherent optical transmission," *IEEE Journal of Lightwave Technology*, vol. 9, no. 2, 1991, pp. 182–193.

[9] Ryu, S. and Y. Horiuchi, "Use of an optical amplifier in a coherent receiver," *IEEE Photonics Technology Letters*," Vol. 3, No. 7, 1991, pp. 663–665.

[10] Kazovsky, L. G. and O. K. Tonguz, "ASK and FSK coherent lightwave systems: A simplified approximate analysis," *IEEE Journal of Lightwave Technology*, vol. 8, no. 3, 1990, pp. 338–352.

[11] Thylen, L., "Integrated optics in LiNbO3: recent developments in devices for telecommunications," *IEEE Journal of Lightwave Technology*, no. 6, 1988, pp. 847–861.

[12] Wood, T. H., "Multiple quantum well waveguide modulators," *IEEE Journal of Lightwave Technology*, no. 6, 1988, pp. 743–757.

[13] Ziemer, R. E., and W. H. Tranter, *Principles of Communications: Systems, Modulation, and Noise*, Houghton Mifflin Co. Boston, 1990.

[14] Goto, M., K. Hironishi, A. Sugata, K. Mori, T. Horimatsu, and M. Sasaki, "A 10 Gb/s optical transmitter module with a monolithically integrated electroabsorption modulator with a DFB laser," *IEEE Photonics Technology Letters*, vol. 2, no. 12, 1990, pp. 896–898.

[15] Ogita, S., Y. Kotaki, M. Matsuda, Y. Kuwahara, H. Onaka, H. Miyata, and H. Ishikawa, "FM response of narrow linewidth, multielectrode quarter wavelength shift DFB laser," *IEEE Photonics Technology Letters.*, vol. 2, no. 3, 1990, pp. 165–166.

[16] Papannareddy, R. "Bit error rate performance analysis of an optical MSK heterodyne/synchronous receiver," *IEEE Global Telecommunications Conf.*, San Diego, Vol. 2, 1990, pp. 503.7.1–503.7.5.

[17] Iwashita, K. and T. Matsumoto, "Modulation and detection characteristics of optical continuous phase transmission systems," *IEEE Journal of Lightwave Technol.*, vol. LT-5, 1987, pp. 452–460.

[18] Wakita, K. et al., "High-speed electro-optic phase modulators using InGaAs/InAlAs multiple quantum well waveguides," *IEEE Photonics Technology Letters*, vol. 1, no. 12, 1989, pp. 441–442.

[19] Kazovsky, L. G., "Performance analysis and laser linewidth requirements for optical PSK heterodyne communications systems," *IEEE Journal of Lightwave Technology*, vol. LT-4, no. 4, 1986, pp. 415–425.

[20] Yamazaki, S, and K. Emura, "Feasibility study on QPSK optical heterodyne systems," *IEEE Journal of Lightwave Technology*, vol. 8, no. 11, 1990, pp. 1646–1653.

[21] Garrett, I. and G. Jacobsen, "Phase noise in weakly coherent systems," IEE Proc., vol. 136, pt. J, no. 3, 1989, pp. 159–165.

[22] Foschini, G. J., L. J. Greenstein, and G. Vannucci, "Noncoherent detection of coherent lightwave signals corrupted by phase noise," *IEEE Trans. Commun.*, vol. 36, no. 3, 1988, pp. 306–314.

[23] Nicolson, G., "Probability of error for optical heterodyne DPSK system with quantum phase noise," *Electron Lett.*, vol. 20, 1984, pp. 1005–1007.

[24] Garrett, I. and G. Jacobsen, "The effect of laser linewidth on coherent optical receivers with nonsynchronous demodulation," *IEEE Journal of Lightwave Technology*, vol. LT-5, 1987, pp. 551–560.

[25] Garrett, I., and G. Jacobsen, "Theoretical analysis of heterodyne optical receivers for transmission systems using semiconductor lasers with non negligible linewidth," *IEEE Journal of Lightwave Technology*, vol. LT-4, no. 3, 1986, pp. 323–334.

[26] Einarsson, E. J. Strandberg, and I. T. Monroy, "Error probability evaluation of optical systems distributed by phase noise and additive noise," *IEEE Journal of Lightwave Technology*, vol. 13, no. 9, 1995, pp. 1847–1852.

[27] Hao, M. and S. Wicker, "Performance evaluation of FSK and CPFSK optical communication systems: A stable and accurate method," *IEEE Journal of Lightwave Technology*, vol. 13, no. 8, 1995, pp. 1613–1623.

[28] Garrett, I. and G. Jacobsen, "Theory for optical heterodyne narrow deviation FSK receivers with delay demodulation," *IEEE Journal of Lightwave Technol*, vol. 6, no. 9, 1988, pp. 1415–1423.

[29] Papannareddy, R., "Linewidth requirements for an optical MSK heterodyne/Synchronous receiver," *IEEE Photonics Technology Letters*, vol. 4, no. 7, 1992, pp. 768–771.

[30] Norimatsu, S. and K. Iwashita, "Linewidth requirements for optical synchronous detection systems with nonlegible loop delay time," *IEEE Journal of Lightwave Technology*, vol. 10, no. 3, 1992, pp. 341–349.

[31] Davis, A. W, M. J. Pettitt, J. P. King, and S. Wright, "Phase diversity technique for coherent optical receivers," *IEEE Journal of Lightwave Technology*, vol. LT-5, no. 4, 1987, pp. 561–572.

[32] Tsao, H. W., J. Wu. S. Yang, and Y. Lee, "Performance analysis of polarization-insensitive phase diversity optical FSK receivers," *IEEE Journal of Lightwave Technology*, vol. 8, no. 3, 1990, pp. 385–395.

[33] Creamer, M. J., et al., "Field demonstration of 565 Mbit/s DPSK coherent transmission system over 176 km of installed fiber," *Electronics Letters*, vol. 24, no. 22, 1988, pp. 710–719.

[34] Walker, N. G., and G. R. Walker, "Polarization control for coherent communications," *IEEE Journal of Lightwave Technology*, vol. 8, no. 3, 1990, pp. 438–457.

[35] Glance, B., "Polarization independent coherent optical receiver," *IEEE Journal of Lightwave Technology*, vol. LT-5, no. 2, 1987, pp. 274–276.

[36] Kavehrad, M. and B. Glance, "Polarization-insensitive FSK optical heterodyne receiver using discriminator demodulation," *IEEE Journal of Lightwave Technology*, vol. 6, no. 9, 1988, pp. 1386–1394.

[37] Ryu, S., S. Yamamoto, Y. Namihira, K. Mochizuki, and H. Wakabayashi, "Polarization diversity techniques for the use of coherent optical fiber submarine cable systems," *IEEE Journal of Lightwave Technology*, vol. 9, 1991, pp. 675–682.

[38] Kazovsky, L. G., "Phase and polarization diversity coherent optical techniques," *IEEE Journal of Lightwave Technology*, vol. 7, no. 2, 1989, pp. 279–292.

[39] Iwashita, K., and N. Takachio, "Chromatic dispersion compensation in coherent optical communications," *IEEE Journal of Lightwave Technology*, vol. 8, no. 3, 1990, pp. 367–375.

[40] Elrafaie, A. F., R. E. Wagner, D. A. Atlas, and D. G. Daut, "Chromatic dispersion limitations in coherent lightwave transmission systems," *IEEE Journal of Lightwave Technology*, vol. 6, no. 5, 1988, pp. 704–709.

[41] Iwashita, K., and N. Takachio, "Experimental evaluation of coherent dispersion distortion in optical CPFSK transmission systems," *IEEE Journal of Lightwave Technology*, vol. 7, no. 10, 1989, pp. 1484–1487.

[42] Winter, J. H., "Equalization in coherent lightwave systems using microwave waveguides," *IEEE Journal of Lightwave Technology*, vol. 7, 1989, pp. 813–815.

[43] Priest, R. G., and T. G. Giallorenzi, "Dispersion compensation in coherent fiber-optic communications," *Optics Lett.*, vol. 12, no. 4, 1984, pp. 179–181.

[44] Alexander, S. B., "Design of wide-band optical heterodyne balanced mixer receivers," *IEEE Journal of Lightwave Technology*, vol. LT-5, no. 4, 1987, pp. 523–537.

[45] Abbas, G. L., V. W. S. Chan, and T. K. Lee, "A dual detector optical heterodyne receiver for local oscillator noise suppression," *IEEE Journal of Lightwave Technology*, vol. LT-3, no. 5, 1985.

[46] Kasper, B. L., et al., "Balanced dual-detector receiver for optical heterodyne communication at gigabit per second rates," *Electronics Letters*, vol. 22, no. 8, 1986.

[47] Agrawal, G. P., "Effect of fiber-far-end reflections in optical communication systems on BER and receiver sensitivity," *IEEE Journal of Lightwave Technology*, no. 1, 1986, pp. 58–63.

[48] Jimlett, J. L., "Effect of interferometric phase-to-intensity-noise conversion by multiple reflections on gigabit/second DFB laser transmission systems," *IEEE Journal of Lightwave Technology*, no. 6, 1989, pp. 888–895.

[49] Clarke, B. L., "Numerical simulation study of performance degradation incurred by high-speed digital optical communication system using intensity modulated laser diodes and p-i-n receivers due to optical reflections," *IEEE Journal of Lightwave Technology*, no. 6, 1991, pp. 741–749.

[50] Chraplyvy, A., "Limitations on lightwave communications imposed by optical fiber nonlinearities," *IEEE Journal of Lightwave Technology*, vol. 8, no. 10, 1990.

[51] Gordon, J. P. and L. F. Mollenauer, "Effects of fiber nonlinearities and amplifier spacing on ultra long distance transmission," *IEEE Journal of Lightwave Technology*, vol. 9, no. 2, 1991, pp. 1170–173.

[52] Watanabe, W. and M. Shirasaki, "Exact compensation for both chromatic dispersion and Kerr effect in a transmission fiber using optical phase conjugation," *IEEE Journal of Lightwave Technology*, vol. 14, no. 3, 1996, pp. 243–248.

[53] Ryu, S., et al., "First sea trial of FSK heterodyne optical transmission systems using polarization diversity," *Electronics Letters*, vol. 24, no. 7, 1988, pp. 399–400.

[54] Imai, T. et al., "Polarization diversity detection performance of 2.5 Gb/s CPFSK regenerators intended for field use," *IEEE Journal of Lightwave Technology*, vol. 9, no. 6, 1991, pp. 761–769.

[55] Ryu, S., et al., "Field demonstration of 195 km long coherent unrepeatered submarine cable system using a booster amplifier," *Electronics Letters*, Vol. 28, No. 21, 1992, pp. 1965–1966.

[56] Gimlett, J. L. et al., "A 2-Gbit/s optical FSK heterodyne transmission experiment using a 1520 nm DFB laser transmitter," *IEEE Journal of Lightwave Technology*, vol. LT-5, no. 9, 1987, pp. 1315–1324.

[57] Gnauck, A. H., et al., "Coherent lightwave transmission at 2 Gb/s over 170 km of optical fiber using phase modulation," *Electronics Letters*, Vol. 23, no. 6, 1987, pp. 286–287.

[58] Saito, S., et al., "An over 2200 km coherent transmission experiment at 2.5 Gb/s using erbium-doped fiber in line amplifiers," *IEEE Journal of Lightwave Technology*, vol. no. 2, 1991, pp. 161–169.

[59] Gnauck, A. H., et al., "4 Gb/s heterodyne transmission experiments using ASK, FSK, and DPSK modulation," *IEEE Photonics Technology Letters*, vol. 2, no. 12, 1990, pp. 908–910.

[60] Vodhanel, R. S., et al., "Performance of directly modulated DFB lasers in 10 Gb/s ASK, FSK, and DPSK lightwave systems," *IEEE Journal of Lightwave Technology*, vol. 8, no. 9, 1990, pp. 1379–1386.

[61] Wagner, R. E., "10 Gb/s modulation of 1.55 μm DFB lasers for heterodyne detection," *in Tech. Dig. 7th Int. Conf. Integrated Optics and Opt. Fiber Commun.*, Japan, 1989, paper, 18c2-1.

[62] Takachio, N., et al., "8 Gb/s 202 km optical CPFSK transmission experiment using 1.3 μm zero dispersion fiber," *Electronics Letters*, vol. 26, no. 8, 1990, pp. 506–508.

[63] Linke, R. A., and R. E. Wagner, "Heterodyne lightwave detection on track towards commercial systems," *IEEE Mag. Lightwave Commun.*, vol. 1, no. 4, 1990, pp. 28–35.

[64] Tonguz, O. K and R. E. Wagner, "Equivalence between premplified direct detection and heterodyne receivers," *IEEE Photonics Technology Letters*, vol. 3, 1991, pp. 835–837.

[65] Green, P. and R. Ramaswami, "Direct detection lightwave systems: why pay more?" *IEEE Mag. Lightwave Commun.*, vol. 1, no. 4, 1990, pp. 36–48.

[66] Heidemann, R., et al., "10 Gb/s transmission and beyond," *Proc. IEEE*, vol. 81, no. 11, 1993, pp. 1558–1567.

Chapter 6

Soliton Lightwave Systems

From the previous chapters, we know that the bit rate–distance product of a lightwave system is mainly limited by the fiber group velocity dispersion and nonlinearity effects. As mentioned in Chapter 2, the nonlinear effect of the fiber (self-phase modulation) can be used to counteract the fiber dispersion, resulting in an undistorted pulse called a soliton. Hence, the generation of a soliton is the result of a balance between the group velocity dispersion and self-phase modulation.

The subject of soliton wave propagation in nonlinear media has been studied in many articles and books [1–8]. First, due to the fiber loss, the decreased peak power of the soliton pulse can lead to soliton broadening, thus requiring periodic soliton amplification. The soliton amplification involves the use of lumped EDFAs or distributed amplification techniques. Optical soliton transmission in optical fibers was first proposed in 1973 [2], and significant progress from theory to reality was accomplished during the 1980s using the stimulated Raman distributed amplification process [7–8]. During the early 1990s, practical soliton lightwave systems have been realized at many research laboratories by utilizing EDFAs as lumped amplifiers.

In recent years, several multigigabit transmission experiments have shown the possibilities of using the soliton systems for very long distance, transoceanic, optical processing and routing networks [9–16]. In comparison with other lightwave systems, soliton lightwave systems must employ the return-to-zero (RZ) format for generating the optical bit stream to minimize soliton interaction.

The main objective of this chapter is to describe the development of single channel soliton lightwave systems, with emphasis on the principles of soliton wave propagation, amplification, system design, and performance degradation issues. Section 6.1 discusses the principles of soliton lightwave systems including soliton wave propagation and the process of soliton amplification. Section 6.2 describes the system design constraints of soliton lightwave systems. The recent field trials and experiments are outlined in Section 6.3. The multichannel soliton lightwave systems are discussed in Chapter 8.

6.1 PRINCIPLES OF SOLITON LIGHTWAVE SYSTEMS

The experimental setup of a typical soliton lightwave system using EDFAs as lumped amplifiers is shown in Figure 6.1. The setup consists of a semiconductor laser source along with an optical isolator, a booster amplifier, an external modulator, lumped EDFAs, and a direct detection receiver. The soliton pulses with durations of a few picoseconds are generated by using the gain switched DFB laser diodes (GS-DFB) [17] followed by a narrow-band optical filter or with the use of monolithic mode-locked external cavity semiconductor (ML-ECL) lasers [18]. In the case of gain switching technique, the DFB laser is used to generate the picosecond optical pulses by biasing the laser below threshold and pumping it high above threshold periodically by applying current pulses. The data pulses are externally modulated by using a lithium niobate Mach-Zehnder intensity modulator, resulting in an RZ optical pulse stream. The lumped EDFAs are pumped at 1.48 μm using multiquantum-well InGaAsP lasers. At the receiving end, the soliton pulses are detected by using a direct detection receiver and a streak camera is utilized to monitor the optical pulse stream. In the following subsections, we discuss the principles of soliton wave propagation and amplification.

6.1.1 Soliton Wave Propagation

The propagation of a soliton in an optical fiber is described by a well-known normalized modified nonlinear Schrodinger equation (NSE) [5]

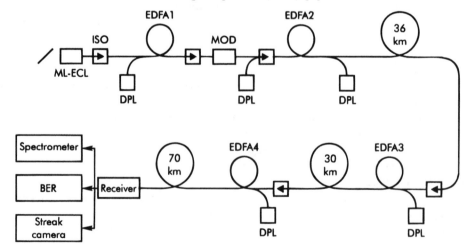

Figure 6.1 Experimental setup of a typical soliton lightwave system. Here, ML-ECL: Mode-locked external semiconductor laser; ISO: Optical isolator; MOD: Modulator; and DPL: Diode pump laser. (*Source:* [10]. © 1990 IEEE. Reprinted with permission.)

$$i\frac{\partial U_n}{\partial Z_n} + \frac{1}{2}\frac{\partial^2 U_n}{\partial T_n^2} + N_s^2|U_n|^2 U_n = -i\Gamma_L U_n + i_R\frac{\partial^3 U_n}{\partial T_n^3} + N_s^2\tau_R U_n\frac{\partial|U_n|^2}{\partial T_n} \quad (6.1)$$

where the normalized electric field, distance, and time (U_n, Z_n, T_n) are defined by

$$U_n = \frac{E}{\sqrt{P_0}} \quad (6.2)$$

$$Z_n = \frac{z}{L_D} \quad (6.3)$$

$$T_n = \frac{t - \dfrac{z}{v_g}}{T_0} \quad (6.4)$$

where E represents the electric field phasor of the soliton pulse, P_0 is its peak power, v_g is the group velocity, and T_0 is the soliton pulse width related to the full width at half maximum (FWHM) of the soliton through $T_0 = T_{FWHM}/1.763$, and L_D is the dispersion length defined as

$$L_D = \frac{T_0^2}{\beta_2} = \frac{2\pi c T_0^2}{\lambda^2|D|} \quad (6.5)$$

where $\beta_2 = -D\lambda^2/2\pi c$ is the group velocity dispersion parameter, with D being the fiber dispersion and λ is the operating wavelength. When $D < 0$, solutions to (6.1) exist in the form of pulses, which are known as bright solitons. On the other hand, when $D > 0$, the solutions are characterized by the absence of intensity in a form of a dip, they are called dark solitons. In this chapter, we consider only the bright solitons. The term N_s is a dimensionless parameter that determines the soliton order defined as [5]

$$N_s = \gamma_p P_0 L_D \quad (6.6)$$

where γ_p is the nonlinearity parameter given by

$$\gamma_p = \frac{2\pi n_2}{\lambda A_e} \quad (6.7)$$

where A_e is effective area of the fiber core and n_2 denotes the Kerr coefficient. The soliton corresponding to $N_s = 1$ is called the fundamental soliton. The solitons corresponding to other integer values of $N_s > 1$ are called higher order solitons. Using

(6.5)–(6.7), the required peak power for a fundamental soliton wave propagation can be written as

$$P_0 = \frac{1}{\gamma_p L_D} = \frac{\lambda^3 D A_e}{4\pi^2 c n_2 T_0^2} \tag{6.8}$$

From (6.8), we can see that the peak power of the soliton pulse mainly depends on the fiber dispersion and the soliton pulse width.

The first term on the R.H.S. of (6.1) represents the fiber loss, characterized by [5]

$$\Gamma_L = \frac{L_D \alpha}{2} = \frac{Z_0 \alpha}{\pi} \tag{6.9}$$

where $Z_0 = \pi L_D / 2$ represents the period of the soliton pulse and α is the fiber loss coefficient. The terms N_s and Z_0 play an important role in the theory of soliton wave propagation. The second term of R.H.S. of (6.1) governs the effect of second order dispersion, characterized by [5]

$$\partial_R = \frac{\beta_3}{6|\beta_2|T_0} \tag{6.10}$$

where β_3 represents the second order dispersion. The third term of the R.H.S. in (6.1) takes into account SRS, where the parameter τ_R is defined by

$$\tau_R = \frac{T_R}{T_0} \tag{6.11}$$

where T_R is related to the slope of Raman gain profile at the carrier frequency. The second and third order terms of R.H.S. in (6.1) need to be included only for the ultrashort optical pulses ($T_0 \ll 10$ ps). One can ignore those terms, provided that the pulse widths are in the range of 10–50 ps.

6.1.2 Soliton Amplification

When the fiber loss term is taken into account, the soliton pulse broadening occurs due to the decrease in peak power. The reduced peak power weakens the nonlinear effect required to counteract the dispersion. Hence, by ignoring the second and third R.H.S. terms, (6.1) can be solved by using the numerical or inverse scattering methods [19,20] by treating Γ_L as a small perturbation. An approximate solution of (6.1) is therefore given by

$$U(Z_n, T_n) = \sec h[\exp(-2Z_n\Gamma_L)T_n] \exp[(i/8\Gamma_L)(1 - \exp(-4Z_n\Gamma_L))] \tag{6.12}$$

When $Z_n = 0$ in (6.12), we get $U_n(0, T_n) = \sec h(T_n)$ as the solution corresponding to the fundamental soliton, implying that the required input pulse shape should be a hyperbolic secant, whose power and pulse width are related by (6.8). In other words, when $N_s = 1$, the effect of fiber dispersion is compensated by the fiber nonlinearity when the input pulse has a hyperbolic secant shape. Nevertheless, other pulse shapes for initial value can also be considered for $N_s = 1$. Figure 6.2 shows the evolution of such an input pulse having a Gaussian shape $U_n(0, T_n) = \exp(-T_n^2/2)$. Note that the Gaussian pulse evolves toward the fundamental soliton by changing its shape, width, and peak power. When N_s is not equal to 1, a similar behavior results. In general, minor deviation from the ideal input hyperbolic secant pulse shape is not the limiting factor for soliton propagation, since the pulse is able to adjust itself to form a fundamental soliton. Also, from (6.12) one can infer that the exponential increase in soliton width can be written in terms of the input soliton width as

$$T_1 = T_0\exp(2Z_n\Gamma_L) \tag{6.13}$$

To overcome the soliton pulse broadening due to the fiber loss, periodic amplification of the soliton, is required along the fiber link. The soliton amplification can be achieved by using two transmission methods: (1) the lumped amplifier method and (2) the distributed amplifier method. The first method utilizes erbium-doped fiber amplifiers to compensate the fiber loss [21–23]. In this method, the path average signal power is different from the power after each amplifier and the launched signal power. Hence, to minimize this energy excursion, the launched signal power has to be adjusted to give a path average power equal to the soliton power.

Additionally, in the case of a large number of cascaded amplifier systems, the accumulation of fiber dispersion can limit the spacing between the two amplifiers.

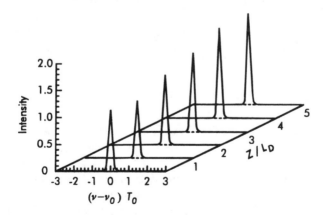

Figure 6.2 Fundamental soliton wave propagation. (*Source:* [5]. Reprinted with the permission of John Wiley and Sons.)

Therefore, the amplifier spacing must be chosen so that the soliton pulse is not perturbed over its length. In one numerical study [24], it has been shown that if the amplifier spacing $L << Z_0$, little or nothing happens to the pulse shape over one amplification period. Then the nonlinear effect accumulated over each L is simply determined by the corresponding path-average power.

Thus, by keeping the path average power constant and equal to the usual soliton power from one period to the next, one can obtain a perfectly behaved soliton. In practical conventional fibers, Z_0 is in the order of 100 km at multigigabit rates, which limits the amplifier spacing to about 20–30 km. However, the value of Z_0 can be increased to 500–1000 km by using dispersion shifted fibers and in this case, the amplifier spacing up to 50 km is feasible.

The second method is the distributed amplifier technique, which can compensate for the fiber loss along the transmission line [25,26]. The distributed amplification can be achieved by the use of stimulated Raman process or rare-earth-doped fibers. The main drawback involved with the stimulated Raman amplification process is that it requires high power semiconductor lasers. The latter method utilizes distributed erbium-doped fibers, in which whole transmission fiber is erbium-doped and the spacing between the pump-power stations is limited to 100 km [26]. The ideal distributed amplifier system has no power excursions. However, signal power excursion may exist due to the use of a finite power, intrinsic loss, non-optimized fiber design, and unequalized spacing.

6.2 SYSTEM DESIGN CONSTRAINTS

The system design constraints are the key factors involved in optimizing the system performance for long distance applications. There are five system design constraints that must be addressed in a soliton lightwave system: (1) amplifier spacing; (2) soliton-to-soliton interaction; (3) timing jitter; (4) frequency chirp; and (5) self frequency shift. These system design constraints are discussed in the following subsections.

6.2.1 Amplifier Spacing

For the lumped amplified soliton lightwave systems, amplifier spacing is an important design parameter in determining the system performance. The amplifier spacing in the case of nonsoliton lightwave systems can be as large as 100 km, but for the soliton lightwave systems, the typical amplifier spacing is around 30 km. In recent years, the lumped amplified soliton lightwave systems have been analyzed in terms of amplifier spacing and the maximum operating bit rate [24,27–29]. The results show that solitons can be propagated over long distances if the amplifier spacing is considerably less than the soliton period.

The large value of amplifier spacing is possible with the use of pre-emphasis method, in which the input peak power is made larger than that of a fundamental soliton for which $N_s > 1$. This technique can be understood by using a model based on average soliton dynamics [24]. In this model, the path average power over a single link is the same as that of the ideal single soliton power. Let us introduce a new variable u, as

$$U_n(Z_n, T_n) = u(Z_n, T_n)\exp(-\Gamma_L Z_n) \tag{6.14}$$

using (6.14) and ignoring the second and third R.H.S. terms in (6.1), we get

$$i\frac{\partial u}{\partial Z_n} + \frac{1}{2}\frac{\partial^2 u}{\partial T_n^2} + N_s^2\exp(-2\Gamma_L Z_n)|u|^2 u = 0 \tag{6.15}$$

Furthermore, if u does not change significantly between the amplifiers, then one can use the average value of N_s defined as

$$N_{\text{avg}} = \frac{N_s^2}{L}\int_0^L \exp(-2\Gamma_L Z_n)\,dz = \frac{N_s^2}{\alpha L}[1 - \exp(-\alpha L)] \tag{6.16}$$

Using the average value of N_s in (6.15), one gets the standard NSE,

$$i\frac{\partial u}{\partial Z_n} + \frac{1}{2}\frac{\partial^2 u}{\partial T_n^2} + N_{\text{avg}}^2|u|^2 u = 0 \tag{6.17}$$

The average value of N_s should be unity for the fundamental soliton wave propagation, and the initial amplitude of the "average $N_s = 1$" soliton is obtained from (6.16) as

$$N_s = \sqrt{\frac{\alpha L}{1 - \exp(-\alpha L)}} \tag{6.18}$$

For example, the required value of N_s is 2.15 for a 100 km amplifier spacing and a fiber loss of 0.2 dB/km. Also, the required value of N_s can be found in terms of the amplifier gain G_0 by using $\exp(-\alpha L) = 1/G_0$ in (6.18). Upon simplification, we get

$$N_s = \sqrt{\frac{G_0 \ln G_0}{G_0 - 1}} \tag{6.19}$$

Assuming $G_0 \gg 1$, (6.19) reduces to $\sqrt{\ln G_0}$. The evolution of an "average $N_s = 1$" soliton over 5,000 km is shown in Figure 6.3 for an initial value of $N_s = 2.15$ with $L = 100$ km, $G_0 = 20$ dB and $Z_0 = 1,000$ km.

In general, the amplifier spacing in soliton lightwave systems is limited by the soliton period Z_0, which in turn depends on the pulse width and group velocity dispersion parameters. The upper limit on the amplifier spacing for a fundamental soliton propagation is often chosen as [24]

$$L < \frac{8}{10} Z_0 \tag{6.20}$$

where $8Z_0$ is the full soliton period. The factor of 10 in (6.20) is arbitrary and chosen to meet the condition that $L \ll 8Z_0$. Equation (6.20) is also known as fiber perturbation criteria. For an $N_s \gg 1$ soliton, the upper limit for L is given by [24]

$$L = \frac{Z_0}{N_s^2} \tag{6.21}$$

Using the value of Z_0 in (6.20), the soliton pulse width can be determined from

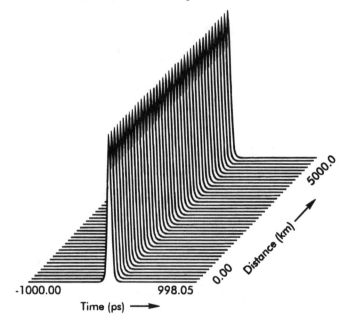

Figure 6.3 Evolution of an "average $N_s = 1$" soliton over 5000 km. (*Source:* [24]. © 1991 IEEE. Reprinted with permission.)

$$T_0 = \sqrt{\frac{10L\lambda^2|D|}{8\pi^2 c}} \qquad (6.22)$$

Equations (6.18)–(6.22) constitute the design rules for choosing the amplifier spacing in lumped amplified soliton lightwave systems.

6.2.2 Soliton-Soliton Interaction

In this section, we consider the interaction of neighboring solitons and the constraint on the maximum bit rate. The soliton interaction can be avoided by keeping the pulses separated by several times their pulse widths. The interaction of solitons in lightwave systems has been analyzed in [30–31]. Let us understand the implications of soliton interaction by solving the NSE by considering the input amplitude consisting of a soliton pair

$$U_n(0, T_n) = \sec h(T_n - \tau_0) + A_r \sec h[A_r(T_n + \tau_0)]e^{j\theta} \qquad (6.23)$$

where the separation between two soliton pulses is represented by $2\tau_0$, A_r is the relative amplitude of the two solitons, and θ is the relative phase. To avoid soliton interaction, the separation between the adjacent solitons should be $2\tau_0 T_0 = T$, where T is the bit period, from which the bit rate can be estimated as

$$B = \frac{1}{2\tau_0 T_0} = \frac{0.88}{\tau_0 T_{\text{FWHM}}} \qquad (6.24)$$

To minimize soliton interaction, the soliton pulse width and soliton separation must be optimized so that the bit rate is maximum for a given distance. However, it is important to note that when $\theta = 0$ (solitons are in phase), the two solitons attract each other such that they collide periodically along the fiber length and for $\theta = \pi$, (solitons are out of phase), the solitons repel each other and their spacing increases with distance. Hence, the soliton interaction also depends on the relative phase of the soliton pair.

Since the soliton separation is determined by the inverse of the data rate, the constraint on the soliton pulse width can be deduced in terms of the maximum bit rate and the total transmission distance by using the condition [27]

$$L_T < \frac{1}{4} Z_P \qquad (6.25)$$

where L_T represents the total transmission distance and Z_P is the periodic collapse length given by [27].

$$Z_P = Z_0 \exp(1/2BT_0) \tag{6.26}$$

Using (6.26) in (6.25) and upon rearranging yields

$$T_0 = \frac{1}{2B \ln\left(\dfrac{4L_T}{Z_0}\right)} \tag{6.27}$$

From (6.27), it can be seen that the maximum soliton pulse width is mainly governed by the bit rate. Hence, for a given data rate and total transmission distance, the maximum permissible soliton pulse width can be determined.

6.2.3 Timing Jitter

The use of optical amplifiers in soliton lightwave systems restore the pulse energy to the original value, and they also produce ASE, which degrades the SNR of the lightwave link. This noise not only degrades the SNR, but also randomly shifts the carrier frequency of the soliton pulse, which leads to changes in the group velocity through dispersive fibers and hence introduces an uncertainty in the arrival time of the optical pulse. A bit error results if the pulse fails to arrive in its assigned time slot. This phenomenon is called the timing jitter, and this effect is analyzed in [32–34] and is commonly called the Gordon-Haus effect.

We use the analysis of [33] to discuss the implications of timing jitter on system design. An expression for the maximum permissible distance can be written in terms of the system parameters as

$$L_m \leq \frac{0.5158}{B} \left[\frac{k_S k_W^2 A_e L A_s}{n_{SP} n_2 Dh[\exp(\alpha L) - 1]} \right]^{1/3} \tag{6.28}$$

where, A_s is the amplitude of the soliton pulse required for a lossy fiber, h is the Planck's constant, n_2 is the nonlinear coefficient, n_{SP} is the amplifier spontaneous emission factor, and the constants k_S and k_W are defined by

$$k_S = BT_{\mathrm{FWHM}} \tag{6.29}$$

$$k_W = BT_W \tag{6.30}$$

where T_W is the half width of the timing window, k_W is restricted by the inequality, $k_W < 0.5$. Equation (6.28) is called the Gordon-Haus limit, which means that the maximum permissible length of a soliton lightwave system is inversely proportional to the data rate and fiber dispersion. The fiber dispersion, however, can be controlled to some extent by selecting the operating wavelength or by choosing the dispersion shifted fibers. Also, it is important to note that the maximum pulse power is directly proportional to the fiber dispersion. Thus, for a given data rate and maximum transmission distance, one can choose the optimum values of power levels and fiber dispersion.

The Gordon-Haus effect can also be minimized by using a band pass filter at the output of each amplifier [35]. The effect of filtering as well the laser linewidth, including the Gordon-Haus effect, has been analyzed in [36]. The results of this analysis is shown in Figure 6.4 for a 10-Gbps soliton lightwave system with and without filters for different laser linewidths.

From Figure 6.4, it can be seen that the optical bandpass filter reduces the timing jitter due to the carrier linewidth and Gordon-Haus effect.

An alternative technique utilizes an in-line optical phase conjugator midway down the system span for compensating the composite dispersive-nonlinear effects. It is shown in [37] that an in-line optical phase conjugation at an optimal point two-thirds of the way down the system span reduces the rms jitter by a factor of three.

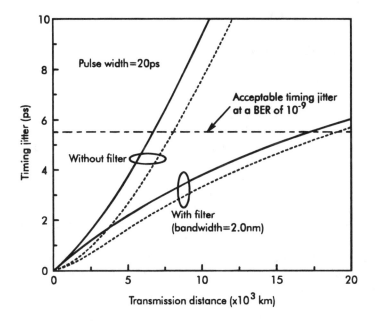

Figure 6.4 Timing jitter of a 10 Gbps soliton lightwave system as a function of transmission distance with and without filters for different laser linewidths. Here, the dotted and solid lines denote the laser linewidths of 1 MHz and 100 MHz, respectively. The parameters used in the calculation are: $D = 0.1$ ps/km/nm, $\lambda = 1.55$ μm, $\alpha = 0.25$ dB/km, $n_2 = 3.18 \times 10^{20}$ m^2/W, $n_{sp} = 1.5$, $A_e = 40$ μm^2, and $L = 30$ km. (*Source:* [36]. © 1995 IEEE. Reprinted with permission.)

Additionally, the effect of timing jitter on the system performance can be reduced without resorting to any in-line soliton controls by using alternating-amplitude solitons and optical time-division multiplexing/demultiplexing techniques [13].

6.2.4 Frequency Chirp

For the soliton wave propagation, the launched pulse should not only a have a hyperbolic secant intensity profile but also should be free from chirp. The effect of chirp on soliton wave propagation has been studied by solving (6.1) with an input amplitude [5]

$$U_n(0, T_n) = \sec h(T_n)\exp(-iC_pT_n^2/2) \tag{6.31}$$

where C_p is the chirp parameter. The analysis has shown that the frequency chirp is not detrimental for $C_p < C_{\text{crit}}$, where C_{crit} is the critical value of C_p, which depends on N_s. For fundamental soliton wave propagation $N_s = 1$, $C_{\text{crit}} = 1.64$. From the system design point of view, the value of C_p should be as small as possible. Low chirp can be obtained by utilizing the monolithically integrated external cavity lasers.

6.2.5 Soliton Self-Frequency Shift

The shift in carrier frequency of the soliton is called the self-frequency shift. A shift in carrier frequency translates into a change in group velocity. This phenomenon can be studied by including the third R.H.S. term in (6.1). For ultrashort optical pulses, the pulse spectrum becomes wider so that the high frequency components can transfer energy through SRS to the low frequency components of the same pulse, which is often called the self-induced SRS or intrapulse SRS [5].

The strength of SRS is dictated by T_R. With the inclusion of this term in (6.1), one can solve (6.1) numerically to analyze the effect of intrapulse SRS on the soliton wave propagation. It has been shown that, due to the intrapulse SRS, the arrival time of soliton becomes a function of the self-frequency shift, which in turn slows down the propagation of a soliton pulse. The intrapulse SRS is a limiting factor for soliton lightwave systems operating at bit rates in excess of 10 Gbps requiring solitons of width 1 ps or less. For bit rates less than 10 Gbps, the impact of the soliton self-frequency shift is negligible.

In conclusion, the maximum bit rate and the transmission distance of a soliton lightwave system are limited by the fiber perturbation, soliton interaction, and timing jitter. Figure 6.5 shows the maximum bit rate versus the transmission distance for different amplifier spacing. From Figure 6.5, it can be seen that the maximum bit rate is limited by the fiber perturbation and soliton interaction for shorter distances and for longer distances, the maximum bit rate is limited by the timing jitter. The

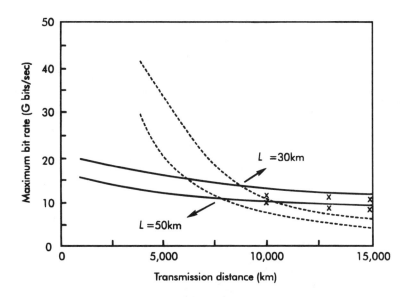

Figure 6.5 Maximum bit rate versus transmission distance for L = 30 and 50 km. The solid line denotes the effect of fiber perturbation and soliton interaction, dotted line represents the effect of timing jitter. (*Source:* [28]. © 1995 IEEE. Reprinted with permission.)

principal limitation for systems operating at higher than 10 Gbps is fiber perturbations resulting from the distributed loss and discrete amplification. This limitation can be overcome by using the dispersion shifted fibers or dispersion compensation techniques.

6.3 SOLITON TRANSMISSION EXPERIMENTS

In recent years, several soliton transmission experiments proved the possibility of realizing the soliton lightwave systems for future very long distance, transoceanic, and optical networks [9–16]. The soliton transmission experiments can be classified into two types: (1) the direct fiber link and (2) the recirculating fiber loop. The experiments using the direct fiber link are more practical than the recirculating fiber loop experiments. In the fiber loop experiment, solitons can be transmitted over several thousands of kilometers. For example, the solitons over 10,000 km can be transmitted by using a 75-km fiber loop and three EDFAs spaced at 25 km apart.

The instability of the optical soliton source was a major problem in many of the experiments, except [11], where a stable optical pulse was generated by using an electro-absorption (EA) modulator and a DFB-laser. In this method, the application of sinusoidal driving voltage to the EA modulator produced an ultra-short stable

Table 6.1
Summary of Soliton Transmission Experiments

Bit rate (Gbps)	Distance (km)	Spacing (km)	System description	Reference
3.2–5	100	25	EDFAs with DSF and GS-DFB lasers	[9]
4	136	36, 30, and 70	EDFAs with CSMF and MESC[a] lasers	[10]
5	6000	30	EDFAs with DSF, EA modulator, DFB lasers, and fiber loop	[11]
8.2	4200	25	EDFAs with DSF, MESC lasers, and fiber loop	[12]
10	12200	30	EDFAs with DSF, GS-DFB lasers, and fiber loop	[13]
20	70	23 and 47	Distributed Raman amplifiers with DSF and GS-DFB lasers	[14]
32	90	30	EDFAs with DSF and MESC lasers	[15]
80	80	20, 30, and 30	EDFAs with DSF and GS-DFB lasers	[16]

[a]MESC:Mode-locked monolitic extended-cavity laser.

optical pulse with the almost hyperbolic secant squared shape. Table 6.1 gives a summary of the recent soliton transmission experiments conducted at 1.55 μm.

Problems

Problem 6.1

Assume a 1.55 μm soliton lightwave system that uses DSF fibers with $D = 0.1$ ps/km-nm. Calculate the peak power required to launch a fundamental soliton into the fiber given $n_2 = 3.2 \times 10^{20}$ m^2/W and $A_e = 40$ μm^2. Also, find the period of the soliton pulse. The soliton pulse shape is assumed to have a hyperbolic secant shape with a FWHM of 20 ps.

Problem 6.2

Find the required value of N_s for a 50 km amplifier spacing, and a fiber loss of 0.2 dB/km. Also, what is the required amplifier gain?

Problem 6.3

Find the required width as well as the period of the soliton pulse in Problem 6.2 given $D = 0.5$ ps/km-nm and $\lambda = 1.55$ μm.

Problem 6.4

Assume a soliton lightwave system required to operate at 10 Gbps and spans a total transmission distance of 10,000 km. What is the required pulse width of the soliton assuming negligible soliton interaction. Assume $Z_0 = 1,000$ km.

Problem 6.5

Calculate the maximum transmission distance of a soliton lightwave system operating at a bit rate of 5 Gbps, if the timing window set by the timing jitter is 20% of the bit duration. Assume $N_s = 2.15$, $n_{SP} = 1.4$, $D = 0.2$ ps/km-nm, $n_2 = 3.2 \times 10^{-20}$ m^2/W and $A_e = 40$ μm^2, $L = 30$ km, $\alpha = 0.2$ dB/km, and FWHM of 20 ps.

References

[1] Mollenaurer, L. F., R. H. Stolen, and J. P. Gordon, *Phys. Rev. Lett.*, vol. 45, 1980, p. 1095.

[2] Hasegawa, A. and F. Tappert, "Transmission of stationary nonlinear optical pulses in dispersive dielectric fibers," *Applied Phys. Lett.*, vol. 23, 1973, pp. 142–143.

[3] Agrawal, G. P. and R. W. Boyd, Eds., *Contemporary Nonlinear Optics*, Academic Press, San Diego, CA 1992.

[4] Dodd, R. K. et al., *"Solitons and Nonlinear Wave Equations*, Academic Press, Orlando, FL, 1984.

[5] Agrawal, G. P., *Fiber-Optic Communication systems*, John Wiley & Sons Inc., New York, 1992, Chapter 9.

[6] Hasegawa, A. and Y. Kodama, "Signal transmission by optical solitons in monomode fiber," *Proc. IEEE*, Vol. 89, no. 9, 1981, pp. 1145–1150.

[7] Mollenaurer, L. F, J. P. Gordon, and M. N. Islam, "Soliton propagation in long fibers with periodically compensated loss," *IEEE J. Quantum Electronics*, vol. QE-22, no. 1, 1986, pp. 157–173.

[8] Mollenaurer, L. F, R. H. Stolen, and J. P. Gordon "Demonstration of soliton transmission over more than 4000 km in fiber with loss periodically compensated by Raman gain," *Opt. Lett.*, vol. 13, 1988, pp. 675–677.

[9] Nakazawa, M., K. Suzuki, and Y. Kimura, "3.2–5 Gb/s 100 km error-free soliton transmissions with erbium amplifiers and repeaters," *IEEE Photonics Technology Letters*, vol. 2, No. 3, 1990, pp. 216–219.

[10] Olsson, N. A., et al., "4 Gb/s soliton data transmission over 136 km using erbium doped fiber amplifiers," *IEEE Photonics Technology Letters*, vol. 2, No. 5, 1990, pp. 358–359.

[11] Taga, H. et al., "Multi-thousand kilometer optical soliton data transmission experiments at 5 Gb/s using an electroabsorption modulator pulse generator," *IEEE Journal of Lightwave Technology*, vol. 12, no. 2, 1994, pp. 231–236.

[12] Hansen, P. B., et al., "8.2 Gb/s, 4200 km soliton data transmission using a semiconductor soliton source," *IEEE Photonics Technology Letters*, vol. 5, No. 10, 1993, pp. 1236–1237.

[13] Suzuki, M. et al., "10 Gb/s over 12200 km soliton data transmission with alternating-amplitude solitons," *IEEE Photonics Technology Letters*, vol. 6, no. 6, 1994, pp. 757–759.

[14] Iwatsuki, K. et al., "20 Gb/s optical soliton data transmission over 70 km using distributed amplifiers," *IEEE Photonics Technology Letters*, vol. 2, No. 12, 1990, pp. 905–907.

[15] Anrekson, P. A., et al., "32 Gb/s optical soliton data transmission over 90 km," *IEEE Journal of Lightwave Technology*, vol. 4, no. 1, 1992, pp. 76–79.

[16] Iwatsuki, K. et al., "80 Gb/s optical soliton transmission over 80 km with time/polarization division multiplexing," *IEEE Photonics Technology Letters*, vol. 5, No. 2, 1993, pp. 245–247.

[17] Nakazawa, M., K. Suzuki, and Y. Kimura, "Transform-limited pulse generation in the GHz region from a gain-switched distributed feedback laser diode using spectral windowing," *Opt. Lett.*, vol. 15, 1990, pp. 715–717.

[18] Wu, M. C., et al., "Tunable monolithic colliding pulse mode-locked quantum-well lasers," *IEEE Photonics Technology Letters*, vol. 3, 1991, pp. 874–876.

[19] Hasegawa, A. and Y. Kodama, "Signal transmission by optical solitons in monomode fiber," *Proc. IEEE*, vol. 69, 1981, p. 1145–1150.

[20] N. J. Doran and K. J. Blow, "Solitons in optical communications," *IEEE Journal Quantum Electronics*, vol. QE-19, 1983, pp. 1883–1888.

[21] Mollenaurer, L. F, S. G. Evangiledes, and H. A. Haus, "Long distance soliton propagation using lumped amplifiers and dispersion shifted fiber," *IEEE Journal of Lightwave Technology*, vol. 9, no. 2, 1991, pp. 194–197.

[22] Smart, R. G., J. L. Zyskind, and D. J. DiGiovanni, "Experimental comparison of 980 nm and 1480 nm pumped saturated in-line erbium-doped fiber amplifiers suitable for long-haul soliton transmission systems," *IEEE Photonics Technology Letters*, vol. 5 no. 7, 1993, pp. 770–772.

[23] Islam, M. N., L. Rahman, and J. R. Simpson, "Special erbium fiber amplifiers for short pulse switching, lasers, and propagation," *IEEE Journal of Lightwave Technology*, vol. 12, no. 11, 1994, pp. 1952–1962.

[24] Blow, K. J. and N. J. Doran, "Average soliton dynamics and the operation of soliton systems with lumped amplifiers," *IEEE Photonics Technology Letters*, vol. 3, No. 4, 1991, pp. 369–371.

[25] Rottwitt, K. et al., "Fundamental design of a distributed erbium doped fiber amplifier for long distance transmission," *IEEE Journal of Lightwave Technology*, vol. 10, no. 11, 1992, pp. 1544–1552.

[26] Rottwitt, K. J. H. Povlsen, and A. Bjarklev, "Long distance transmission through distributed erbium-doped fibers," *IEEE Journal of Lightwave Technology*, vol. 11, no. 12, 1993, pp. 2105–2115.

[27] Evans, A. F. and J. V. Wright, "Constraints on the design of single channel (>10 Gb/s) soliton systems," *IEEE Photonics Technology Letters*, vol. 7, no. 1, 1995, pp. 117–119.

[28] Chi, S., J.-C. Dung, and S. Wen, "The maximum bit rates of soliton communication systems with lumped amplifiers and filters in different distances," *IEEE Journal of Lightwave Technology*, vol. 13, no. 6, 1995, pp. 1121–1126.

[29] Knox, F. M., W. Forysiak, and N. J. Doran, "10 Gbit/s soliton communication systems over standard fiber at 1.55 μm and the use of dispersion compensation," *IEEE Journal of Lightwave Technology*, vol. 13, no. 10, 1995, pp. 1955–1962.

[30] Kodama, Y. and K. Nozaki, "Soliton interaction in optical fibers," *Opt. Letts.*, vol. 12, no. 12, 1987, pp. 1038–1040.

[31] Kodama, Y. and S. Wabnitz, "Reduction of soliton interaction forces by bandwidth limited amplification," *Electronics Letters*, vol. 27, 1991, pp. 1931–1933.

[32] Gordon, J. P. and A. H. Haus, "Random walk of coherently amplified solitons in optical fiber transmission," *Opt. Lett.*, vol. 11, 1986, pp. 665–667.

[33] Marcuse, D. "An alternative deviation of the Gordon-Haus effect," *IEEE Journal of Lightwave Technology*, vol. 10, no. 2, 1992, pp. 273–278.

[34] Ding, M., and K. Kikuchi, "Limits of long-distance soliton transmission in optical fibers with laser diodes as pulse sources," *IEEE Photonics Technology Letters*, vol. 4, No. 6, 1992, pp. 667–669.

[35] Kodama, Y. and A. Hasegawa, "Generation of asymptotically stable solitons and suppression of the Gordon-Haus effect," *Opt. Lett.*, vol. 17, no. 1, 1992.

[36] Iwatsuki, K. et al., "Timing jitter due to carrier linewidth of laser-diode pulse sources in ultra-speed soliton transmission," *IEEE Journal of Lightwave Technology*, vol. 13, no. 4, 1995, pp. 639–649.

[37] Forysiak, F. and N. J. Doran, "Reduction of Gordon-Haus jitter in soliton transmission systems by optical phase conjugation," *IEEE Journal of Lightwave Technology*, vol. 13, no. 5, 1995, pp. 850–855.

Chapter 7

Optoelectronic Integrated Circuits

The high-speed performance of lightwave systems using discrete devices and interconnects is limited by the parasitic elements that are introduced at very high frequencies. These parasitic elements can be minimized by using optoelectronic integrated circuits. Optoelectronic integrated circuits (OEICs) involve the integration of electronic and optical devices and optical interconnects. The integration of active and passive devices will lead into complex switching, modulation, and signal processing capabilities. The advancement of material research plays a critical role in optoelectronic integration, and much of the work in that area is still under development.

In recent years, the subject of OEICs has been discussed in several review articles [1–8] and books [9–10]. The objective of this chapter is to discuss the principles of OEICs with the emphasis on recent developments. Section 7.1 describes the principles of OEICs and the integration of guided wave devices is discussed in Section 7.2. Integrated lightwave transmitters are discussed in Section 7.3, and Section 7.4 describes the integrated lightwave receivers. Finally, the OEIC arrays are discussed in Section 7.5.

7.1 PRINCIPLES OF OPTOELECTRONIC INTEGRATED CIRCUITS

Figure 7.1 shows the basic elements of an OEIC. It consists of an integrated transmitter, an in-line amplifier, a receiver, and the optical interconnects. In this circuit, several functions such as light transmission, amplification, combined with detection are performed. Thus, all active and passive components are integrated to form a complete lightwave system. Such a system is called an OEIC. However, the development of an OEIC requires different materials, different device structures, and various processing steps (unlike in silicon ICs, where many of the components are made with the same material and same processing steps).

Optoelectronic integration can be classified into three types: (1) hybrid; (2) monolithic; and (3) flip-chip. In the case of hybrid integration, the discrete devices

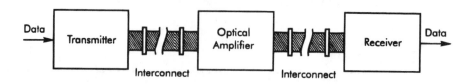

Figure 7.1 Basic elements of an OEIC.

on separate functional blocks are connected using electronic leads or optical fiber interconnects. For example, the laser diode can be integrated with a driver circuit, which consists of a bipolar transistor to form a transmitter, or a photodetector can be integrated with a low noise front-end amplifier to form a receiver. Additionally, it offers flexibility towards the selection of a material for any given function, such as GaAs for a laser, Si for electronic circuitry, and a polymer for a mechanical structure. The inherent disadvantage of this scheme is the lack of compactness and enhanced parasitic effects due to bonding, lead wires, and interconnects.

In monolithic integration, all the active and passive components are fabricated on the same chip, giving rise to less parasitic effects and yields robust performance as well as high-speed for lightwave transmission and receiving systems. In contrast with the monolithic integration, the flip-chip concept enables the integrated receiver to be fabricated from state-of-the-art GaAs or InGaAs or silicon IC technologies. Using the flip-chip technology, both optical and electronic components can be independently optimized, and the parasitic effects are expected to be small. In practice, the preferred method of integration depends on the system application and the cost. In the following subsections, we will explore these technologies with specific examples.

7.1.1 Hybrid Integration

Figure 7.2 shows a schematic example of a hybrid optoelectronic transreceiver, which uses a III–V based laser transmitter and receiver mounted on a silicon substrate through the use of V-grooves, waveguides, and optical interconnects between the active chips and fibers. For a system shown in Figure 7.2, the optimal technologies are employed for the transmitter, receiver, and substrate. In addition, good thermal and physical properties of silicon have proven to be advantageous with regard to robust and reliability.

7.1.2 Monolithic Integration

The monolithic integration, like electrical ICs, yields structures that are aligned photolithographically to offer compactness, high-reliability, and robust performance against vibration or shock. However, the technology required to fabricate different components on a single wafer can be complex, and isolation problems may arise.

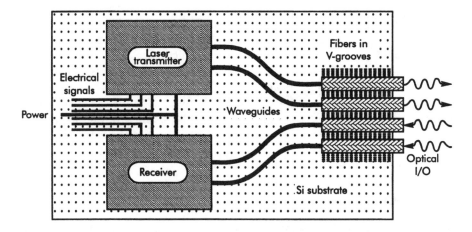

Figure 7.2 A hybrid Si/III-V based transreceiver module. (*Source:* [9]. Reprinted with permission.)

The schematic diagram of a DBR laser monolithically integrated with a Mach-Zehnder interference modulator is shown in Figure 7.3. Using the monolithic integration, the alignment and coupling problems that arise with the bulk-optical connections are eliminated. The Mach-Zehnder interference device is used as an external modulator. The operation of this modulator is discussed in Section 7.2. Similarly, an optical amplifier may also be monolithically integrated with a photodiode [11].

The use of monolithic integration is beneficial in WDM systems, in which several signals of different wavelengths can be modulated individually and are sent over a single mode fiber [12,13]. In this scheme, 16 channels of different wavelengths

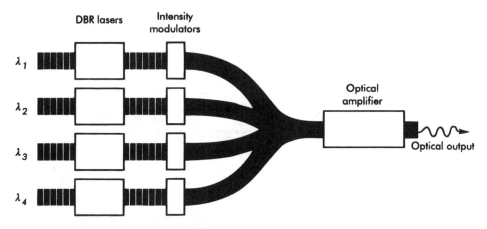

Figure 7.3 Monolithic integration of a DBR laser with a Mach-Zehnder interference modulator. Here, λ_1, λ_2, λ_3, and λ_4 denote the wavelengths of different signals. (*Source:* [9]. Reprinted with permission.)

from 1.544 to 1.553 μm with an average channel spacing of 0.6 nm are multiplexed into a single output using 16 DBR lasers and electroabsorption modulators on a 4 × 6 μm chip [12]. For data transmission applications, monolithic integration allows the fabrication of lasers of closely spaced wavelengths. It also simplifies the alignment of lasers to modulators and mirrors to detectors, as well as minimizes the external coupling requirement to a single mode fiber. This type of monolithically integrated scheme is expected to become popular in the near future. High volume combined with low cost consumer optoelectronic applications such as "fiber to the home" or individual customer terminals will stimulate the rapid development of this technology.

7.1.3 Flip-Chip Integration

Figure 7.4 illustrates the flip-chip integration structure of a receiver. In contrast with the monolithic integration, flip-chip integration enables the integrated receiver to be fabricated from state-of-the-art InGaAs photodetectors and GaAs or Si IC technologies. Using flip-chip technology, both electronic and optical elements can be independently optimized. In the above receiver structure, PIN photodiode has a photosensitive area of 17 μm in diameter and exhibits a low 20-fF capacitance with a quantum efficiency of 80%. The sensitivity of this receiver structure is about −28.5 dBm at a bit rate of 1.8 Gbps.

7.1.4 Materials and Processing Techniques

The OEIC technologies can be broadly classified into two types: (1) GaAs-based OEICs that uses GaAs substrate for short wavelength systems and (2) InP-based

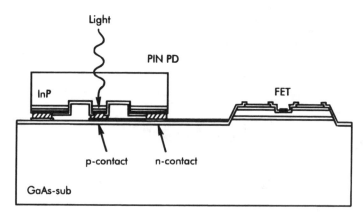

Figure 7.4 Flip-chip integration structure of the receiver. (*Source:* [14]. © 1988 IEE. Reprinted with permission.)

OEICs that use InP as a substrate for long wavelength systems. Another approach towards the fabrication of long wavelength OEICs is heteroepitaxy or the use of mismatched materials that include III-V or II-VI compounds or GaAs and InP-based compounds on Si. This new approach combines the advantages of InP and GaAs devices with the mature Si-process technology. Hence, the choice of materials and processing techniques mainly depends on the operating wavelength, lattice matching considerations, and type of components.

The first step in the development of an OEIC involves the integration of optical and electronic elements on GaAs or InP substrates, preserving their original device structures. The GaAs-based OEICs are utilized in the areas of local area networks, computer interconnects, and optical information processing, whereas the InP-based OEICs are applicable for high capacity long-haul systems. At present, the techniques used for processing OEICs include epitaxial and photolithography [10]. It is expected that special processing techniques will be developed for the fabrication of OEICs in the near future.

7.2 OEIC GUIDED WAVE DEVICES

The guided wave devices perform a variety of functions such as directional coupling, filtering, and modulation. These devices are monolithically integrated with other optoelectronic devices or electronic devices. The guided wave devices can be broadly classified into two types: (1) active devices and (2) passive devices. The active devices include lasers, directional couplers, and electro-optic modulators. Examples of passive devices are lenses, mirrors, and gratings for beam focusing, reflection, and filtering respectively. Additionally there are quasi-passive or active devices that can be integrated. In the following paragraphs, we discuss some of the OEIC guided wave devices.

7.2.1 Modulators

As discussed in Chapter 1, the amplitude of the laser source can be modulated by an external device called a modulator. The modulation can be accomplished by using a wide variety of structures: (1) electro-optic modulators; (2) interferometers; and (3) quantum-well modulators. The operation of electro-optic modulators is based on the principle of the electro-optic effect (the application of an electric filed changes the refractive index of the device, resulting in amplitude or phase variations of the signal output). In recent years, these have traditionally been designed using the large electro-optic coefficient material $LiNbO_3$ (lithium niobate), but for use in OEICs, these devices are being made with semiconductors.

The operation of an interferometer modulator is based on the electro-optic effect. The most popular guided wave modulator/switching device is the Mach-Zehnder interferometer, which is shown in Figure 7.5. In this scheme, with no applied bias, the incoming signal is equally split between the two branches of interferometer,

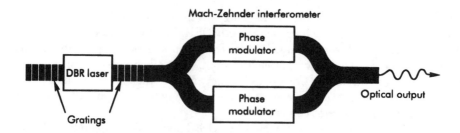

Figure 7.5 Schematic diagram of a guide wave Mach-Zehnder interferometer. (*Source:* [9]. Reprinted with permission.)

without any phase shift and hence the signals are recombined at the other end, yielding maximum intensity output. If the phase difference between the two branches of the interferometer is 180 degrees (can be achieved by suitable bias), the modes in the two branches are 180 degree out of phase, which results in minimum power output. Thus, the intensity of the output signal can be controlled by varying the voltage. Using this interferometer, amplitude or phase modulation can easily be achieved.

Another type of modulator based on the multiple quantum-well structure is shown in Figure 7.6. The principle of operation of this modulator is based on the quantum confined Stark effect [9]. The change in absorption with applied voltage leads to strong polarization dependent electro-absorption which is known as quantum-confined Stark effect. Without quantum wells, the structure would, of course, operate as a bulk electro-optic modulator. The relative amounts of amplitude and phase

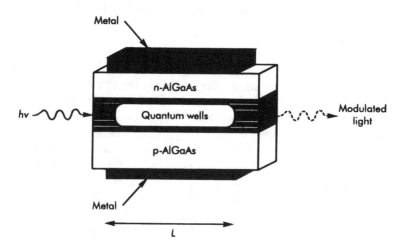

Figure 7.6 A waveguide-based multi-quantum-well modulator. (*Source:* [9]. Reprinted with permission.)

modulation are a function of the quantum-well width, shape, barrier height, and operating wavelength. Optimization of layer structures has been done for the GaAs/AlGaAs as well as for InGaAsP/InGaAs/InP systems [15]. Additionally, the polarization independence has been achieved by using the parabolically graded wells or stain in the InGaAs/InGaAlAs [16,17].

7.2.2 Optical Interconnects

The rapid advances in OEICs coupled with the progress in multiprocessor-based computing and signal processing are placing ever increasing demands on the interconnect medium through which high-speed signals can be both routed and switched [18]. The ultimate speed of the system is not limited by the speed of the individual chips, but by the interconnect delays. The optical interconnects can be broadly classified into three types: (1) inter-chip; (2) intra-chip; and (3) inter-board. The inter-chips are used in an OEIC, whereas the intra-chips are utilized between the chips, and the inter-board interconnects are employed between the two printed circuit boards. The interconnects are required between a group of devices that can perform calculations, share data, and perform input-output operations in high-speed LANs. In addition to LAN applications, they are also used in long-haul, switching, broadband integrated service digital network (B-ISDN), and optical computing systems. At present, the optical interconnect technology is still under development and some of the interface/coupling problems need to be resolved.

7.3 OEIC TRANSMITTERS

A typical lightwave transmitter circuit consists of a light source and driver electronics. Hence, the light source needs to be integrated with the associated driver electronics. A typical 5-Gbps monolithically-integrated 1.5-μm lightwave transmitter using multiple quantum well laser and HBT (heterojunction bipolar transistor) driver circuit is shown in Figure 7.7. Figure 7.7(a) illustrates the cross-sectional view of an OEIC transmitter. The laser-HBT wafers were grown by metalorganic vapor phase epitaxy (MOVPE) technique. The active region of the laser source consists of three 50-Angstroms-thick, compressively strained, InGaAs quantum wells with 200-Angstroms-thick InGaAsP barriers. The active region is surrounded each side by two step index confinement layers, each having a thickness of 100 Angstroms. The p-type InP ridge guide mesa is 3 μm wide and 1.5 μm thick. The n-contact metal and the laser stripe are laterally separated by only 15 μm to minimize the series resistance. The HBT is of the single hetereostructure type with a nominal base thickness of 1,000 Angstroms and a collector thickness of 1,000 Angstroms. The laser and HBT mesas are etched down to the semi-insulating substrate and laterally buried in polymide to minimize any parasitic capacitances. Lines of Cr/Au are printed on the top of the wafer for electrical interconnections between the discrete devices.

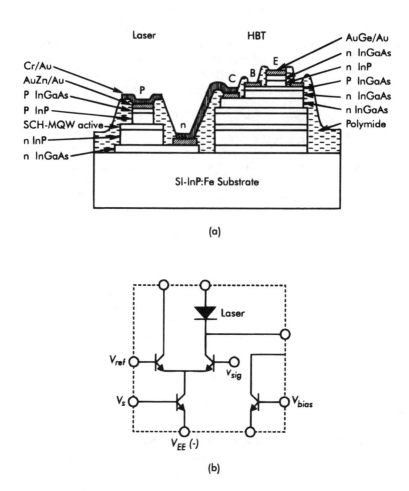

Figure 7.7 (a) Schematic cross-sectional view of the OEIC transmitter and (b) schematic circuit diagram of a laser transmitter. (*Source:* [19]. © 1991 IEEE. Reprinted with permission.)

Figure 7.7(b) shows a schematic diagram of the OEIC transmitter. It consists of a laser and four HBTs ($Q_1 - Q_4$). The laser modulation circuit consists of two HBTs (Q_1 and Q_2) as an emitter-coupled differential pair and a third HBT (Q_3) as a current source. The current supplied by Q_1 is routed by the differential pair. A fourth HBT Q_4 is used for dc biasing the laser. The peak modulation current of the transmitter is set by V_S, and the modulation signal is applied to V_{sig} relative to V_{ref}. The size of the OEIC chip is 900×310 μm and the threshold currents of 18–25 mA have been obtained with a power output a few mW at 5 Gbps. The other reported OEIC transmitters are compared in Table 7.1.

Table 7.1
Comparison of OEIC Transmitters

Technology	Bit rate	Threshold Current	Power output	Reference
InGaAsP-strained				
DFB-MESFET	10 Gbps	90 mA	4.6 dBm	[20]
InGaAsP/InP				
DFB/MODFET	10 Gbps	8 mA	3 mW	[21]
InGaAsP/InP				
EAM/DFB	10 Gbps	—	−1 dBm	[22]
InGaAsP-InP				
MQW/HBT	5.0 Gbps	20 mA	3 mW	[19]

Note: MESFET: Metal semiconductor FET; MODFET: Modulation doped FET; EAM: Electroabsorption modulator.

7.4 OEIC RECEIVERS

Recall from Chapter 4, that an optical receiver consists of a photodetector, a front-end amplifier, an equalizer, and data decision circuits. The integration of the photodetector and the front-end amplifier is the key element in determining the receiver performance. Monolithically integrated photoreceivers consisting of photodetector and preamplifier circuits have been developed over the recent years, because they offer advantages relating to high-speed operation, compactness, high reliability, and packaging simplicity.

Figure 7.8 shows a monolithically integrated photoreceiver using an InAlAs/InGaAs PIN-HBT-based transimpedance amplifier. The receiver structure shown in Figure 7.8(a) is grown by molecular beam epitaxy (MBF) on semi-insulating InP substrate. The different layers used and their dimensions are shown in the Figure. Figure 7.8(b) shows the circuit diagram of the receiver. It consists of a 10 μm × 10 μm PIN diode and a 5 μm × 5 μm emitter HBT amplifier in a transimpedance configuration with a 300-Ohm feedback resistor and a buffer stage that provides 50-Ohm matching output impedance. The receiver circuit requires only two bias voltage supplies, one for the amplifier and the other for the PIN diode. The receiver has demonstrated a bandwidth of 7.1 GHz. The experimental results pertaining to the other reported OEIC receiver are compared in Table 7.2.

7.5 OEIC ARRAYS

Another important goal of the monolithical integration scheme is the development of multiple optoelectronic devices or arrays for WDM systems. An array consists of

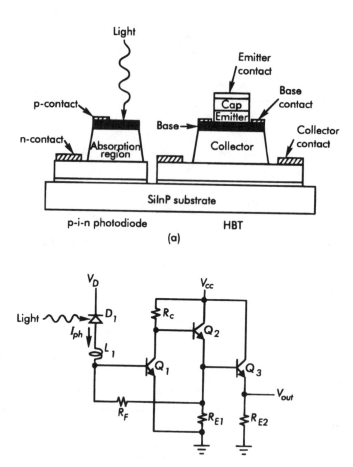

Figure 7.8 (a) Cross sectional view of a PIN-HBT receiver and (b) circuit diagram of the integrated PIN-HBT receiver. (*Source:* [23]. © 1994 IEEE. Reprinted with permission.)

an assemblage of identical devices monolithically integrated on the same chip. The arrays would be useful in a system where a large number of fibers are terminated, for example, at central office terminals where large numbers of fibers are converging from remote locations. Arrays can serve a variety of applications. The arrays of light sources or transmitters can provide higher power output with single longitudinal mode of operation [31]. Wavelength tunability is also possible using a two-dimensional array of sources. Similarly, photodetector arrays or multichannel photoreceiver OEICs [32,33–35] are very attractive for application in optical interconnection, high-speed optical transmission, and wavelength division multiplexing systems due to their

Table 7.2
Comparison of OEIC Receivers

Technology	Bit rate	Sensitivity	3 dB BW	Type	Reference
InAlAs/InGaAs-Inp-MSM-HEMT	3 Gbps	—	3 GHz	TZ-DD	[24]
InGaAs balanced PIN and InP-InGaAs cm HBT	200 Mbps	−49 dBm	3 GHz	Coherent heterodyne	[25]
InGaAs PIN and InAlAs/InGaAs HEMT	10 Gbps	−16.5 dBm	8 GHz	TZ-DD	[26]
WGPD and InAlAs-InGaAs HEMT	10 Gbps	−15.7 dBm	8.3 GHz	TZ-DD	[27]
All-Silicon	10 Gbps	−29.3 dBm	8.2 GHz	TZ-DD	[28]
PIN and InP-InGaAs HBT	12 Gbps	−17.6 dBm	8.2 GHz	TZ-DD	[29]
PIN and InP-InGaAs HBT	20 Gbps	−17.0 dBm	10.4 GHz	TZ-DD	[30]

Note: MSM: Metal semiconductor metal; HEMT: High electron mobility transistor; WGPD: Waveguide photodetector; TZ-DD: Transimpedance-direct detection.

inherent advantages of compactness and packaging simplicity. Arrays are also useful in parallel operation of computers. It is important to note that the development of arrays is still in its infancy, and the problems relating to impedance matching, electrical and optical interference and material growth and compatibility have to be solved.

References

[1] Koch, T. L., and U. Koren, "Semiconductor Photonic Integrated Circuits," *IEEE Journal of Quantum Electronics*, Vol. 27, 1991, pp. 641–653.

[2] Koren, U., "Optoelectronic integrated circuits," in *Optoelectronic Technology and Lightwave Communications Systems*, (Lin, C. Ed.), New York: Van Nostrand Reinhold, 1989, pp. 677–694.

[3] Forrest, S. R., "Monolithic optoelectronic integration: a new component technology for lightwave communications," *IEEE Journal of Lightwave Technology*, Vol. LT-30, 1985, pp. 1248–1264.

[4] Horimatsu, T. and M. Sasaki, "OEIC technology and its applications to subscriber loops," *IEEE Journal of Lightwave Technology*, Vol. 7, No. 11, 1989, pp. 1612–1622.

[5] Bar-chaim, N., et al., "GaAs integrated optoelectronics," *IEEE J. Electron Devices*, Vol. ED-29, 1982, pp. 1372–1392.

[6] Ichino, H., et al., "Over 10 Gb/s IC's for future lightwave communications," *IEEE Journal of Lightwave Technology*, Vol. LT-12, No. 2, 1994, pp. 308–319.

[7] Pedrotti, K. D., et al. "High-bandwidth OEIC receivers using bipolar transistors: design and demonstration," *IEEE Journal of Lightwave Technology*, Vol. 11, No. 10, 1993, pp. 1601–1614.

[8] Das, M. B., J. W. Chen, and E. John, "Designing optoelectronic integrated circuit receivers for high sensitivity and maximally flat response," *IEEE Journal of Lightwave Technology*, Vol. 13, No. 9, 1995, pp. 1876–1884.

[9] Zappe, H. P., *Introduction to semiconductor integrated optics*, ARTECH House Inc., Norwood, MA, 1995, Chap. 12.

[10] Bhattacharya, C., *Semiconductor Optoelectronic Devices*, Prentice-Hall, Inc., New Jersey, 1994, Chap. 12.

[11] Liou, K. Y., et al., "Operation of integrated InGaAsP-InP optical amplifier-monitoring detector with feedback control circuit," *IEEE Photonics Technology Letters*, Vol. 2, 1990, pp. 878–880.

[12] Young, M. G., et al., "A 16 × 1 wavelength division multiplexer with integrated distributed bragg reflector lasers and electroabsorption modulators," *IEEE Photonics Technology Letters*, Vol. 5, 1993, pp. 908–910.

[13] Matz, R., et al., "Development of a photonic integrated transreceiver chip for WDM transmission," *IEEE Photonics Technology Letters*, Vol. 6, 1994, pp. 1327–1329.

[14] Makiuchi, M., "GaInAs PIN photodiode/GaAs preamplifier photoreceiver for gigabit-rate communication systems using flip-chip bonding techniques," *Electronics Letters*, Vol. 24, 1988, pp. 995–996.

[15] O'Brien, et al., "Monolithic integration of an AlGaAs laser and an intracavity electroabsorption modulator using selective partial interdiffusion," *Applied PhySics Letters*, Vol. 58, 1991, pp. 1363–1365.

[16] Inoue, H., et al., "Field-induced refractive index change dependence on incident light wavelength and polarization to an InGaAs/InAlAs multi-quantum-well," *Integrated Photonics Research*, New Orleans, LA, 1992, pp. 338–339.

[17] Zucker, J. E., et al., "Interferometric quantum well modulators with gain," *IEEE Journal of Lightwave Technology*, Vol. 10, 1992, pp. 924–932.

[18] Special issue on optical interconnections for information processing, *IEEE Lightwave Technology*, Vol. 13, No. 6, 1995.

[19] Liou, K. Y., et al. "A 5 Gb/s monolithically integrated lightwave transmitter with 1.5 μm multiple quantum well laser and HBT driver circuit," *IEEE Photonics Technology Letters*, Vol. 3, No. 10, 1991, pp. 928–930.

[20] Miyamoto, Y. et al., "10 Gb/s strained MQW DFB-LD transmitter and superlattice APD receiver module using GaAs MESFET ICs," *IEEE Journal of Lightwave Technology*, Vol. 12, No. 2, 1994, pp. 332–342.

[21] Lo, Y. H., "Multigigabit/s 1.5 μm λ/4 shifted DFB OEIC transmitter and its use in transmission experiments," *IEEE Photonics Technology Letters*, Vol. 2, No. 9, 1990, pp. 673–674.

[22] Goto, M., "A 10 Gb/s optical transmitter module with a monolithically integrated electroabsorption modulator with a DFB laser," *IEEE Photonics Technology Letters*, Vol. 2, No. 12, 1990, pp. 896–898.

[23] Cowles, J. et al., "7.1 GHz bandwidth monolithically integrated InGaAs/InAlAs PIN-HBT transimpedance photoreceiver," *IEEE Photonics Technology Letters*, Vol. 6, No. 8, 1994, pp. 963–965.

[24] Chang, G. K., "A 3 GHz transimpedance OEIC receiver for 1.3–1.5 μm fiber optic systems," *IEEE Photonics Technology Letters*, Vol. 2, No. 12, 1990, pp. 896–898.

[25] Chandrasekhar, S., et al., "Monolithic balanced PIN/HBT photoreceiver for coherent optical heterodyne communications," *IEEE Photonics Technology Letters*, Vol. 3, No. 6, 1991, pp. 537–539.

[26] Akahori, Y., et al., "10 Gb/s high-speed monolithically integrated photoreceiver using InGaAs PIN PD and planar doped InAlAs/InGaAs HEMTs," *IEEE Photonics Technology Letters*, Vol. 4, No. 7, 1992, pp. 754–755.

[27] Muramoto, Y., et al., "High-speed monolithic receiver OEIC consisting of a waveguide PIN photodiode and HEMTs," *IEEE Photonics Technology Letters*, Vol. 7, No. 6, 1995, pp. 685–687.

[28] Tezuka, H., et al., "All-silicon IC 10 Gb//s optical receiver," *IEEE Photonics Technology Letters*, Vol. 4, No. 7, 1992, pp. 754–755.

[29] Lunardi, L. M., et al., "A 12 Gb/s high performance, high sensitivity monolithic PIN/HBT photoreceiver module for long wavelength transmission systems," *IEEE Photonics Technology Letters*, Vol. 7, No. 2, 1995, pp. 182–184.

[30] Lunardi, L. M., et al., "20 Gb/s monolithic PIN/HBT photoreceiver module for 1.55 μm applications," *IEEE Photonics Technology Letters*, Vol. 7, No. 10, 1995, pp. 1201–1203.

[31] Wada, O., et al., "Optoelectronic integrated four-channel transmitter array incorporating AlGaAs/ GaAs quantum well lasers," *IEEE Journal of Lightwave Technology*, Vol. 7, No. 1, 1989, pp. 186–197.

[32] Govindarajan, M., et al., "DC to 2.5 Gb/s × 4 PIN/HBT optical receiver array with low crosstalk," *IEEE Photonics Technology Letters*, Vol. 5, No. 12, 1993, pp. 1397–1399.

[33] Chandrasekhar, S., et al., "Eight channel PIN/HBT monolithic receiver array at 2.5 Gb/s channel for WDM applications," *IEEE Photonics Technology Letters*, Vol. 6, No. 10, 1994, pp. 1216–1218.

[34] Yano, H., et al., "5 Gb/s four-channel receiver optoelectronic integrated circuit array for long-wavelength lightwave systems," *Electronics Letters*, Vol. 28, No. 5, 1992, pp. 503–504.

[35] Takahata, K., et al., "10 Gb/s two-channel monolithic photoreceiver array using waveguide PIN Pds and HEMTs," *IEEE Photonics Technology Letters*, Vol. 8, No. 4, 1996, pp. 563–564.

Chapter 8

Multichannel Lightwave Systems

The low loss single mode optical fiber provides a bandwidth of more than 30 THz over 1.2 to 1.3 μm wavelengths. An information capacity of 30 Tbps can deliver several hundreds or even thousands of channels on a single mode fiber. Hence, the large bandwidth of the fiber can be utilized by transmitting several channels simultaneously on a single mode fiber via multiplexing techniques, which leads to multichannel systems. Multichannel systems provide savings in the costs of installing and upgrading lightwave systems. The use of optical amplifiers in multichannel systems can further increase the capacity of the system. Multichannel systems have proven to be very attractive for LANs, metropolitan networks, and long-haul trunk applications. It is expected that future networks will have to satisfy the needs of large and diverse groups of markets, including traditional voice and data, interactive and broadcast video, fields of health and education, and business and entertainment.

The subject of multichannel systems has been addressed in the literature [1–6]. The objective of this chapter is to describe the different types of multichannel lightwave systems, including their basic principles, system design issues, and practical implementation aspects. Section 8.1 discusses the basic principles of multichannel systems and different system configurations. Section 8.2 describes wavelength division multiplexing (WDM) systems, and the WDM devices are discussed in Section 8.3. The subcarrier multiplexing systems (SCM) are described in Section 8.4, and Section 8.5 gives a perspective on fiber in the loop systems. Finally, Section 8.6 outlines the system design and performance degradation issues.

8.1 PRINCIPLES OF MULTICHANNEL SYSTEMS

In a multichannel system, the signal from each channel is transmitted by using multiplexing techniques. The four main techniques used for signal multiplexing are: (1) optical time-division multiplexing (OTDM); (2) WDM or optical frequency division multiplexing (OFDM); (3) subcarrier multiplexing (SCM); and (4) optical code division multiplexing (OCDM).

In the OTDM technique, the data stream at a higher bit-rate is generated directly by multiplexing several lower bit rate TDM channels utilizing a single laser source [4]. In this scheme, the multiplexing is carried out on the optical signal and electrical to optical conversion is done before multiplexing. On the contrary, the electrical time-division multiplexing scheme is purely a digital technique and signal multiplexing is carried out in the electrical domain to obtain different digital hierarchies. The synchronous optical network (SONET) standard has been utilized to define the synchronous frame structure for transmitting TDM digital signals. The basic level of the SONET hierarchy has a bit rate of 51.84 Mbps, which is referred to as OC-1, where OC stands for optical carrier. Higher hierarchy levels results in a bit rate that is an exact multiple of the basic level. For example, the bit rate for the OC-48 level corresponds to 2.488 Gbps.

The WDM technique refers to the transmission of multiple carriers, individually modulated by independent data bit streams [6]. The wavelength spacing between the carriers is on the order of 1 nm, and in the OFDM scheme, the frequency spacing is on the order of the signal bit rate. Except for the spacing, we, therefore, make no distinction between WDM and OFDM. In recent years, the WDM systems have proven to be very attractive for high capacity and multiple access communication systems [1,3].

The third configuration is based on subcarrier multiplexing (SCM) technique. In this scheme, several analog or digital signals are multiplexed by using microwave subcarriers rather than the optical carriers. The multiplexed signal is combined and then transmitted using a single optical carrier. The different microwave carriers are demultiplexed at the receiver using the conventional microwave techniques. The SCM systems have attracted significant attention due to its application for video distribution by the cable TV industry [2].

The last multiplexing configuration OCDM is based on the spread spectrum technique called the code-division multiplexing (CDM) scheme [5]. In this scheme, each channel is coded in such a way that its spectrum spreads over a wider region than that occupied by the original signal. The multiplexed signal is decoded by the receiver based on a prior knowledge of the code used at the transmitter.

In practice, OTDM systems are limited in their performance due to bottlenecks in device technologies and fiber dispersion [4]. On the other hand, OCDM systems, even though receiving considerable attention [5], are still under research and development. Therefore, in the following sections, we look at the performance and design issues of WDM and SCM multichannel systems, including their implementation aspects.

8.2 WAVELENGTH DIVISION MULTIPLEXING SYSTEMS

Commercially available local area networks with either bus or ring topologies usually provide data rates ranging from 10-Mbps Ethernet to 100-Mbps fiber distributed data interface (FDDI). These networks, however, suffer from electronic bottlenecks (the electronics at each node cannot operate at higher data rates). Additionally, these topologies are inadequate to provide the broadband services which are shown in Figure 8.1. Hence, future networks will have to provide multiple concurrent high bit rate channels. One way to achieve this in an optical network is by allocating different wavelengths for each channel, which is commonly referred to as WDM.

The WDM systems have been the subject of research and development during recent years [6–20]. Systems operating in the 1.3–1.55 μm wavelength regions have been developed for multichannel transmission. Figure 8.2 shows two typical WDM system configurations. In a WDM system, the modulated carrier signals are combined to yield a single multiplexed output. At the receiving end, the multiplexed signal is demultiplexed by using frequency selective components such as a receiving tuned filter or local oscillator. The receivers can be either direct detection or coherent detection type. In the case of unidirectional transmission as shown in Figure 8.2(a), several channels of different wavelengths can be transmitted in one direction on a single mode fiber, whereas Figure 8.2(b) shows the case for the transmission of signals in two directions at different wavelengths.

The current trend of WDM technology is to have minimum spacing between the channels, which is referred to as dense WDM or high density WDM [6]. Dense WDM, with channel spacing between 1 and 10 nm, can accommodate 10–100 optical channels per fiber without coherent detection techniques. Additionally, the use of OEIC transmitter and receiver arrays in WDM systems can further increase system capacity. One of the advantages of WDM is that the use of tunable transmitters and/or receivers on the nodes allows greater network flexibility as channels can be dynamically allocated according to traffic requirements. In the following subsections,

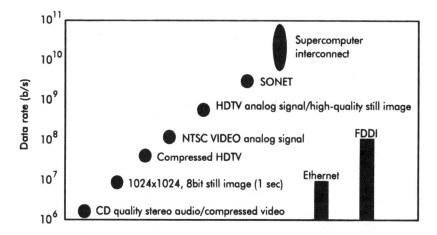

Figure 8.1 Bit rate required for broadband services. (*Source:* [7] © 1992 IEEE. Reprinted with permission.)

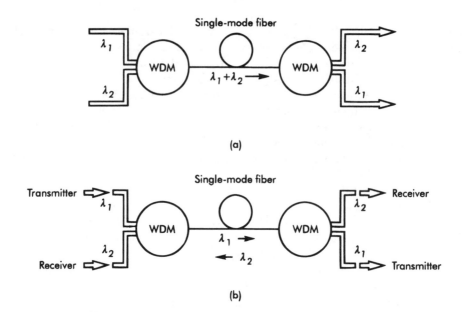

Figure 8.2 WDM system configurations: (a) unidirectional transmission and (b) bidirectional transmission.

we discuss the WDM system architectures, including their performance and device technologies.

8.2.1 WDM Architectures

The WDM systems can be broadly classified into three general architectures: (1) point-to-point links; (2) broadcast-and-select networks; and (3) LANs. First let us look at the point-to-point architecture for long-haul systems. Figure 8.3 depicts the architecture of a point-to-point WDM system. In this architecture, the signal output at each wavelength $\lambda_1, \lambda_2, \ldots, \lambda_N$ from the transmitters is multiplexed together and the demultiplexer routes each channel to its own receiver. If N channels at bit rates $B_1, B_2, \ldots,$ and B_N are multiplexed and transmitted over a fiber of length L, then the capacity of the point-to-point system is $BL = (B_1 + B_2 + \cdots + B_N)L$.

Next, consider the architecture of a broadcast-and-select network as shown in Figure 8.4. In this network, all inputs are combined in a star coupler and broadcast to all outputs. This type of architecture is quite useful for the transmission of several TV channels to a subscriber. The subscriber receives all the channels and the receiver can be tuned to select one of the channels. The capacity of a broadcast-and-select network is mainly limited by the distribution loss and insertion loss of the system. However, the optical amplifiers can be used to compensate these losses.

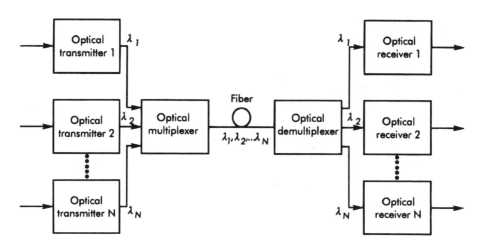

Figure 8.3 Point-to-point WDM system.

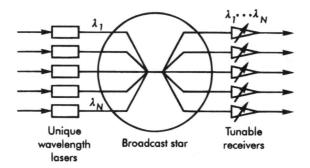

Figure 8.4 The broadcast-and-select WDM system with fixed wavelength lasers and tunable receivers. (*Source:* [6] © 1990 IEEE. Reprinted with permission.)

The third architecture of WDM system is the multiple-access LANs, which is different from broadcast-and-select networks. LANs provide bidirectional access to each subscriber. For example, subscriber-loop or local-loop networks can provide telephone services, or LANs can be used for networking multiple computers. The WDM local area networks have many topologies whose architectures, mainly depend on the type of switching or routing and distribution techniques [7–15]. One of these architectures is shown in Figures 8.5. The LAMBDANET architecture consists of an $N \times N$ star coupler, which distributes the signal to every receiving node of the network. LAMBDANET makes use of an array of N receivers at each receiving node in the network, using a grating demultiplexer to separate the different optical channels. Each receiving node simultaneously receives all of the traffic of the entire network. This feature creates an internally passive, nonblocking, and completely connected

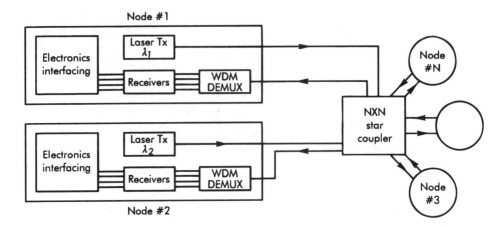

Figure 8.5 The architecture of a LAMBDANET, a hybrid WDM system. (*Source:* [8] © 1990 IEEE. Reprinted with permission.)

network whose capacity and connectivity can be controlled electronically at its periphery. Due to the passive nature of the fiber, grating demultiplexer, and the star coupler, the LAMBDANET is transparent to the bit rate or signal format. Hence, different subscribers can transmit data at different bit rates with different signal formats. This network is useful for the transmission of voice-traffic in a telephone interoffice network or for video broadcast services.

Another LAN architecture called STARNET, a new broadband optical network based on coherent detection technology, is shown in Figure 8.6. It offers both a moderate speed packet switch network and a high-speed broadband circuit interconnect [9]. The bit rate of the packet-switched network is 100 Mbps and the bit rate for the broadband circuit switched network is 3 Gbps per station. The architecture of this network is based on a passive star topology. At every node, the transmitter transmits two independent data streams, broadband circuit data, and the packet

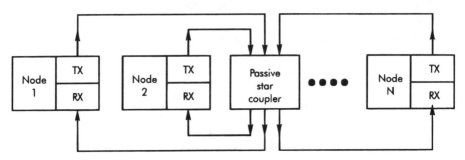

Figure 8.6 The architecture of a STARNET. (*Source:* [9]. © 1993 IEEE. Reprinted with permission.)

data. The tunable receiver at each node can be tuned to any transmitter, thus enabling a broadband circuit interconnect between all the nodes. Additionally, every node is equipped with a fixed receiver that recovers the packet data stream of the previous node.

Other architectures such as fast optical cross connect (FOX) [10], hybrid packet switching system (HYPASS) [11], and passive photonic loop (PPL) [12] have been investigated in recent years for multichannel applications. The FOX scheme is based on the fast tunable transmitter lasers and fixed receivers in an application to parallel processing computers. The main objective of the FOX scheme was to provide a high-speed cross connect between multiple processors and multiple shared memory units on a computer network. In the case of a HYPASS scheme, it uses both tunable transmitters and receivers and hybrid electronics and optics. It is mainly designed for the transmission of high speed packets. The PPL scheme, also sometimes known as passive optical network (PON), is mainly designed for the transmission of voice traffic in the subscriber loop, where a central office routes a signal to the residential customers. These PONs offer an attractive scheme for bringing the fiber to the home (FTTH) or fiber to the curb (FTTC) services.

8.2.2 WDM Network Experiments

During the recent years, there has been steady transition of WDM network experiments from concepts to field trials [7–15]. These experiments can be categorized based on the type of data transmission and detection schemes. The following paragraphs are devoted to the description of WDM metropolitan area networks and LANs using direct and coherent detection schemes and soliton pulses.

8.2.2.1 Direct Detection WDM Systems

The direct detection WDM systems use direct detection optical receivers with tunable optical filters for channel selection. The direct detection receivers offer simplicity, low cost and minimum complexity. The block diagram of a typical direct detection WDM receiving system is shown in Figure 8.7. The direct detection receiving system shown in Figure 8.7(a) consists of a tunable Fabry-Perot filter (FFP), an isolator, and the conventional receiver circuit. The structure of a tunable FFP is shown in Figure 8.7(b) whose transmission characteristics are depicted in Figure 8.7(c). It can be seen that the transmission characteristics of the filter is periodic with transmission peaks spaced by the free spectral range (FSR). As illustrated in Figure 8.7(c), the number of channels that can fit in one FSR is $(N - 0.5)f_c$, where f_c is the minimum channel spacing. The bandwidth of the filter B is measured by the FSR/F ratio, where F denotes the finesse factor. The FSR can be adjusted by varying the spacing L between the mirrors of FFP. The channel selection obtained by the tunable FFP filter is shown in Figure 8.7(d).

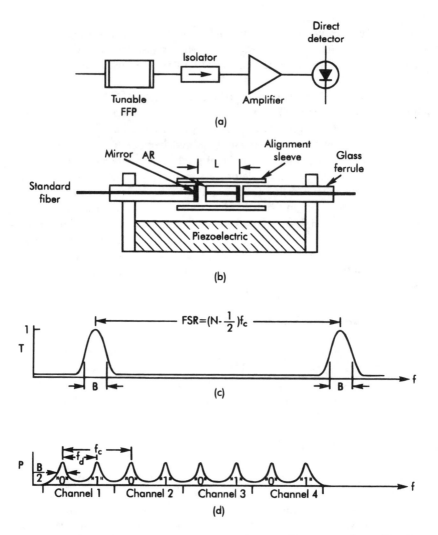

Figure 8.7 Direct detection WDM receiving system: (a) Fabry-Perot filter receiver; (b) tunable Fabry-Perot filter; (c) tunable Fabry-Perot transmission; and (d) channel selection. (*Source:* [21]. © 1990 IEEE. Reprinted with permission.)

The direct detection WDM systems include TeraNet and RAINBOW network experiments [13,14]. The TeraNet was developed to study all seven layers of the open systems interconnection (OSI) standard, and it is a hybrid multiple access scheme which combines WDM and SCM techniques. The network provides either 1-Gbps ATM packet-switched or 1-Gbps circuit-switched access based on a passive star topology. The transmitters use DFB lasers and each laser is assigned a unique wavelength and a subcarrier multiplexed frequency. The channels are spaced by 1.5 nm and each wavelength supports four to six subcarrier channels. The receivers use Fabry-Perot tunable filters to select different wavelengths. The subcarriers are selected by electronic filtering.

The RAINBOW network is a metropolitan area network (MAN) designed to cover a diameter of 55 km. It connects 32 IBM PS/2s, through a 32 × 32 passive star coupler and allows the computers to communicate circuit-switched data at a rate of 200 Mbps node. Each computer is equipped with its own fixed frequency transmitter and a tunable receiver. The transmitters utilize DFB laser diodes and the channel spacing is approximately 1.6 nm. The wavelength selection is achieved by using a tunable Fabry-Perot filter at the receiver.

8.2.2.2 Coherent WDM Systems

The direct detection WDM systems suffer from tighter power budgets and channel spacing, and increased tuning speeds or new network architectures. The use of a coherent detection scheme alleviates the problems of tighter power budgets and channel spacing at the cost of system complexity. The coherent detection scheme consists of a tunable laser along with a balanced receiver circuit.

The coherent WDM system experiments include UCOL (ultra wideband coherent optical network) [15] and STARNET [9]. The UCOL consists of network interface units/access control units that communicate on 20 WDM channels over a passive star topology. The user can access each channel through a time division multiplexing access mode. This system supports data rates starting from a fraction of a megabits per second to 155 Mbps. The channel spacing is about 3.6 GHz. The transmitters use external cavity lasers, tunable over 1 nm in the 1.55 μm wavelength window.

The STARNET is a new coherent WDM network, developed at Stanford University. It is used for high-speed packet switching as well as for broadband interconnect services. The initial experiment connects four workstations through a 4 × 4 passive star coupler. The network operates at a center wavelength of 1.319 μm with a diameter of 4 km. Each node of the network consists of a transmitter, a 3-Gbps broadband receiver for the circuit interconnect, and a 125-Mbps receiver for the packet-switched network. The ultra-linewidth Nd:YAG lasers were used for the transmitters and receivers. The 3-Gbps circuit switch data is PSK modulated on the optical carrier and 100-Mbps packet data after encoding is ASK modulated on the optical carrier at 125-Mbps. The channel spacing of the system is 8 GHz.

8.2.2.3 Soliton WDM Systems

Recall from Chapter 6 that the capacity of a single channel lightwave system can be increased by using a soliton pulse transmission. Furthermore, multiple soliton pulses can be multiplexed to transmit many channels over a single fiber. In recent years, WDM technology has been utilized in soliton systems for achieving higher system capacities particularly useful for transoceanic distances [18–20,22]. However, in a soliton WDM system, the solitons on one channel periodically collide with the neighboring channels due to nonlinear interaction between the solitons. This causes severe interchannel interference, leading to collision-induced timing jitter [18]. The collision-induced timing jitter can be minimized by making the solitons overlap over a distance considerably longer than the amplifier spacing. This can be measured by defining a collision length as the distance over which one soliton propagates over the full width half maximum of the other soliton. If the collision length is about twice the amplifier spacing, then the effect of soliton collision is negligible.

The WDM network experiments, which were discussed in the previous paragraphs, are summarized in Table 8.1.

Table 8.1
WDM Network Experiments

Network	Bit rate	Format	Type of receiver	Transmitters per node	Receivers per node	Reference
Teranet	1 Gbps	ASK	DD	2 (fixed)	2 (tunable)	[13]
Rainbow	200 Mbps	ASK	DD	1 (fixed)	1 (tunable)	[14]
Lambdanet	1.5 Gbps	ASK	DD	1 (fixed)	1 (tunable)	[8]
Starnet	3 Gbps	PSK	Coherent	1 (fixed)	1 (fixed) 1 (tunable)	[9]
Ucol	155 Mbps	DPSK	Coherent	1 (tunable)	1 (tunable)	[15]

8.2.3 WDM Devices

The realization of WDM transmitting and receiving systems requires tunable transmitters and/or receivers. The tunable transmitters can be realized by using tunable lasers. On the other hand, the tunable receivers can be implemented by placing a tunable filter before the receiver in the case of direct detection systems, whereas tunable laser local oscillators are used in coherent detection systems. The other devices required for WDM systems are star couplers and optical amplifiers. These devices have been discussed in the previous chapters. In the following subsections, we discuss the device technologies associated with tunable lasers, tunable filters, and the multiplexers/demultiplexers.

8.2.3.1 Tunable Lasers

The lasers can be tuned by using a tunable element inside the laser cavity. The tunable lasers can be classified into two types: (1) external cavity lasers and

(2) integrated lasers. Depending on the tuning method, the laser tuning can be either continuous or discrete over a particular range. External cavity lasers provide low linewidths and large tuning ranges with moderate speeds [23]. For external cavity lasers, the tuning methods include acousto-optical as well as electro-optical. However, these tuning methods require precise alignment between the laser and the external cavity. In the case of integrated tunable lasers, they can be tuned at very high speeds, but at the expense of tuning range and linewidth. The integrated tuned lasers include DFB/DBR lasers and a three section laser, utilizing an integrated vertical coupler filter (VCF) [24–26].

For coherent WDM systems, the single frequency tunable laser local oscillators are used to achieve the channel selection [27–29]. However, these tunable laser local oscillators are limited by tuning speed and frequency stability. Additionally, coherent receivers require polarization control and frequency stabilization. The polarization maintaining fibers or polarization diversity receivers can be utilized to match the polarization of the received signal with that of the local oscillator. Absolute frequency references have been achieved in both the 1.3–1.5 μm windows [30,31]. Table 8.2 gives a summary of the tuned laser characteristics.

Table 8.2
Tunable Laser Characteristics

Technology	Tuning range	Type of tuning	Linewidth	Speed	Reference
Acoust-optics	70 nm	discrete	–	3 μs	[23]
Electro-optics	7 nm	discrete	60 kHz	100 μs	[23]
3-Section DFB	2 nm	continuous	500 kHz	–	[26]
3-Section DBR	4.4 nm	continuous	1.9 MHz	10 μs	[26]
3-Section VCF	57 nm	discrete	–	–	[25]

8.2.3.2 Tunable Filters

The tunable filters can be classified into three categories: (1) passive; (2) active; and (3) semiconductor laser amplifiers [21,32,33]. The passive filters are basically the wavelength selective components, which are tunable by varying some mechanical element of the filter such as mirror position or etalon angle. This includes tunable Fabry-Perot filters and tunable Mach-Zehnder integrated interferometer filters. In the case of FP filters, the channel or wavelength selection is related to the value of the finesse F of the filter. The typical FP filters have a finesse of up to 200, which implies a number of channels of 30 for the case of FSK DD systems. The number of channels can be increased by increasing the F by cascading the FP filters. The FP filters offer fine frequency resolution but are primarily limited by their tuning speed and losses. In contrast, the Mach-Zehnder interferometer tunable filter is an integrated waveguide device with $\log_2 N$ stages, each of which passes every other incoming wavelength in a divide-by-two fashion until only the desired wavelength remains. This type of filter can separate 100 wavelengths with an optical spacing of 10 GHz. The number of channel selections is limited by the number of stages required and the corresponding losses.

In active category, the filters are based on wavelength-selective polarization by either electro-optic or acousticoptic methods. In the electroptic case, the wavelength selection can be done by varying the dc voltage on the electrodes. On the other hand, the wavelength selection is by changing the acoustic drive frequency in the case of acoustic optic method. Both types of filters can be controlled electronically and have reasonable tuning speeds.

The third category of tuned filters is based on semiconductor laser amplifiers. The DFB laser amplifiers can be tuned by simply adjusting the injection current, but results in change in gain and bandwidth. An improved version of the DFB tunable filter includes a phase control section, which allows separate adjustment of the gain and wavelength, yielding constant gain and bandwidth [33]. Table 8.3 summarizes the tunable filter characteristics.

Table 8.3
Tunable Filter Characteristics

Technology	Tuning range	Tuning Speed	Loss (dB)	Channels	Reference
FP filter	50 nm	ms	5	1000 (two stages)	[32]
MZ-waveguide	45 A	ms	–	128	[32]
Acoustic-optics	400 nm	ms	100s	5	[32]
Electro-optics	10 nm	ns	10	5	[32]
DFB laser amplifier	1–4 nm	ns	10	0	[32]

8.2.3.3 Multiplexers/Demultiplexers

The function of wavelength multiplexers is to combine the output of several transmitter outputs, whereas the demultiplexer splits the received multichannel output into individual channels. The simplest type of demultiplexers are based on the dispersive element, such as a prism or a diffraction grating, which disperses the incident light into several wavelengths. The other type is based on Mach-Zehnder integrated waveguide. In both cases, the same device can be used as a multiplexer or a demultiplexer, depending on the direction of propagation. It is important to note that the demultiplexers require wavelength selective components and the multiplexers can be designed without them. Figure 8.8 shows the configuration of a 10 GHz spaced silica based integrated optic Mach-Zehnder eight channel multiplexer/demultiplexer. In this configuration, seven MZ interferometers and seven phase shifters made of thin-film

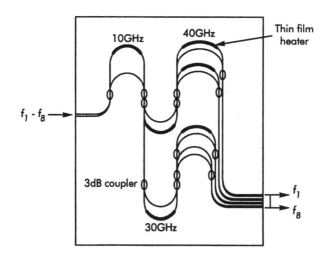

Figure 8.8 Configuration of a 10-GHz spaced silica based integrated optic Mach-Zehnder eight channel multiplexer/demultiplexer. (*Source:* [34]. © 1990 IEEE. Reprinted with permission.)

heaters are integrated on one chip. Each of the eight multiplexed signals is obtained from all eight output ports and in reverse eight signals with 10 GHz spacing can be multiplexed to one waveguide.

Comparison with the MZ configuration, in grating type, the multichannel signal is focused onto a reflection grating which splits various components and a lens focuses them into individual components. However the use of a graded-index lens simplifies alignment.

8.3 SUBCARRIER MULTIPLEXING SYSTEMS

Figure 8.9 shows the basic block diagram of a subcarrier multiplexed system. The analog or digital signals are modulated with microwave subcarriers and then they are combined by using a combiner. The combined signal is then modulated by intensity modulation or external modulation techniques. The modulated lightwave signal is transmitted over a single mode fiber and the lightwave signal is detected by a direct detection or coherent detection techniques. Finally, the baseband signal is recovered by the down conversion by using a microwave receiver.

The SCM techniques are attractive to transmit a very large number of channels for broadband services [35–47]. Services such as cable television, interactive video, video telephony, video-on-demand, high definition television can be handled by using SCM techniques. The SCM techniques offer several advantages over WDM systems. First, the SCM systems use a single laser for transmission and may not require

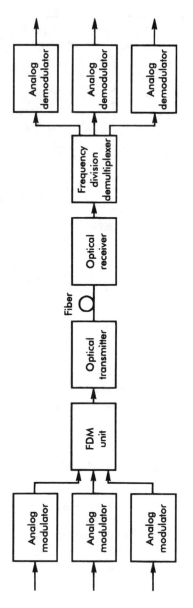

Figure 8.9 Block diagram of a subcarrier multiplexed system.

frequency stabilization. Second, both analog and digital signals can be transmitted (channels 1 to 5 are assigned for analog signals and channels 6 to 10 represent digital signals) in one system. Third, commercially available electronics can be utilized in SCM systems, resulting in low cost. Furthermore, SCM techniques can be used in combination with passive optical networks for building low cost broadband networks. The SCM networks provide broadband services to large number of subscribers from a single transmission point (feeder). The number of subscribers served from a single feeder can be further enhanced by using in-line optical amplifiers. In the following subsections, we describe the different SCM system configurations and their experiments.

8.3.1 IM/Direct Detection SCM Systems

The detection of digital signals by using a direct detection receiver was discussed in Chapter 4. The detection of analog signals is also similar to that of electrical AM or FM receivers except for the quantum noise. The modulated optical signal in an analog SCM system can be expressed by the following expression [35]

$$P(t) = P_0 \left[1 + \sum_{i=1}^{n} m_i \cos(\omega_i t + \varphi_i) \right] \quad (8.1)$$

where N denotes the number of channels, and P_0 denotes the average received power. The terms m_i, ω_i, and φ_i denote the optical modulation index, angular frequency, and phase of the ith channel respectively. The performance of analog SCM systems is measured by the carrier to noise ratio (CNR) instead of SNR, as the two are interrelated, which depends on the laser intensity noise (RIN) as well as on the intermodulation product noise (IMP). The IMP noise results from the mixing of subcarriers due to the nonlinearity of the laser. The CNR requirements of SCM system depend on the modulation scheme [40,41]. For commercial broadcast AM systems, the required weighted SNR is 56 dB and is 22.5 dB for commercial broadcast FM systems.

8.3.2 Coherent Detection SCM Systems

As we learned from the previous Section, the CNR requirement for analog IM/DD SCM systems is quite large, particularly for AM systems. On the other hand, the use of coherent detection techniques in SCM systems can offer improvement in receiver sensitivity [44–47]. Additionally, the use of digital modulation of subcarriers would improve signal quality with a low CNR requirement. In this case the system CNR is degraded by the interchannel crosstalk and IMP. The required value of CNR generally depends on the modulation scheme, which is typically 16–20 dB. The FSK format is commonly used for modulating microwave subcarriers.

8.3.3 SCM Multiple-Access Networks

The SCM systems can be used in conjunction with WDM techniques to provide integrated broadband services i.e., audio, video, and data to large number of subscribers [48–50]. The configuration of such a scheme is shown in Figure 8.10. The configuration shown in Figure 8.10 is used for local area networks. Station # i receives the predetermined signals of the subcarrier f_i and sends any one of the subcarrier signals f_n where $n = 1$ to N. The scheme offers the flexibility of mixing the analog or digital signals by using different subcarriers. Additionally, EDFAs can be used to overcome the losses in multiple-access local area networks and multichannel video distribution systems.

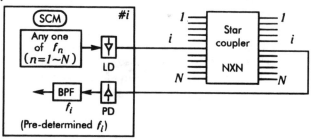

Figure 8.10 SCM multiple-access system configuration. (*Source:* [77]. Reprinted with permission.)

8.3.4 SCM System Experiments

During recent years, several laboratory experiments have demonstrated the potential of SCM systems. In the case of analog SCM systems, both AM and FM modulation formats have been used. In [40], a 60-channel FM video experiment was carried out using the SCM scheme. Similarly, in another AM experiment, 35 video channels were transmitted with 6-MHz channel spacing, by utilizing microwave subcarriers [41]. In addition, the recent experiments have demonstrated the use of optical amplifiers in analog SCM systems [42,43]. The use of EDFAs have increased the power budget of the SCM systems by compensating the distribution losses.

Furthermore, several coherent SCM experiments were demonstrated using ASK, FSK, QPSK, and differential PSK formats [44–48]. In [44], a 20-channel FSK system was built using differential detection. The 20 subscribers span the 2–6 GHz band with 200-MHz channel spacing. In [46], 8 Gbps SCM was designed with a QPSK format. Additionally, ten 50-Mbps video channels were transmitted using differential QPSK format on five microwave subcarriers [47]. Also, a two-channel ASK/FM experiment at 560 Mbps has been verified using SCM techniques [48].

The SCM systems described in the previous paragraphs are point-to-point transmission experiments. In recent years, several multiple-access SCM networks have been demonstrated [49–52]. Passive splitting of video signals to 16–32 subscribers has been achieved over a SCM passive optical network [49]. Also, optically amplified SCM distribution networks have been demonstrated [50]. Furthermore, a cost-effective broadband passive optical network (PON) architecture that incorporates WDM, TDM, and SCM access has been demonstrated for multichannel transmission [51,52]. The PON provides 39 Mbps downstream to each of the optical network units (ONU) for the transmission of compressed digital video channels.

8.4 FIBER IN THE LOOP SYSTEMS

The fiber in the loop (FITL) systems are expected to be deployed in telecommunication networks both by telephone and cable television companies, which provide lower operating costs and high-bandwidth services into the 21st century. Today, fiber is often the clear economic choice over copper based networks for serving large business, backbone feeder, and interoffice applications. The telephone companies have designed a number of fiber optic subscriber loop systems to deliver both voice and video services. On the other hand, the cable TV companies have deployed the fiber for delivering video to residential customers. These fiber links to the customers will play a critical role in the future introduction of broadband integrated services digital network (BISDN), in which high speed services are feasible.

The telecommunication network is comprised of trunk systems and subscriber loop systems. The trunk systems connect central offices (CO), whereas the subscriber loop systems connect central offices with subscribers. The subscriber loop system is also called a fiber in the loop (FITL) system, which can be classified into several categories, depending on the loop architecture. Several FITL architectures are shown in Figure 8.11. In FTTC architecture as shown in Figure 8.11(a), the multiplexed signals are transmitted between the CO and the optical network units (ONUs). The ONU contains an optical transmitter and receiver, including the multiplexing/demultiplexing devices. It is cost-effective for providing today's video and/or plain old telephone services (POTs) over a fiber. In a FTTC system, fiber is brought only as far as the service access points (SAP) or pedestals. The metallic copper drops carry POTs and video signals from the SAP to several residences. Several SAPs can communicate with the CO over the same fiber through TDM or other techniques.

In contrast, fiber to the home (FTTH) architectures extend the fiber all the way to the customer premises. Figure 8.11(b) illustrates a simple FTTH architecture. A passive $1:N$ optical power splitter located at the remote node divides downstream optical signals roughly equally among the N subscribers, and also combines light transmitted upstream among the N subscribers. The passive nature of the link between the CO and customer premises has given rise to the term called passive optical network (PON). The FTTH architecture eliminates the metallic cables and the availability of low-cost transmitters and receivers facilitates the installation of ONUs at individual customer premises rather than at shared SAP locations.

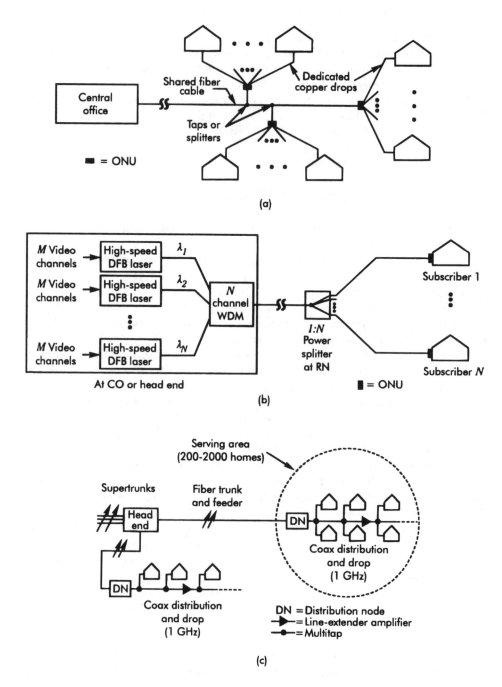

Figure 8.11 FITL architectures: (a) Fiber to the curb (FTTC); (b) Fiber to the home (FTTH); (c) CATV fiber/coax distribution. (*Source:* [53,78]. © 1989, 1993 IEEE. Reprinted with permission.)

Figure 8.11(c) shows another architecture for cable TV transmission. This scheme evolved from the existing coaxial-cable systems. In this approach, fiber "feeders" are used to deliver a large number of analog and digitized video signals to distribution nodes located near clusters of 200–2,000 customers. From the distribution nodes, the video signals are distributed over existing coaxial cables. This scheme results in a smaller number of repeaters, higher coaxial bandwidth, and reduced maintenance problems. Using this architecture, several hundred TV channels can reach each of several hundred customers, thus providing every customer with individualized programming.

8.5 MULTICHANNEL SYSTEM DESIGN ISSUES

The design factors and degradation issues play an important role in determining the performance of multichannel systems. The design factors involved in a WDM system are channel spacing, filter or multiplexer design, and frequency stabilization. In the case of SCM systems, the required CNR or SNR, number of channels, channel spacing, intermodulation distortion, and signal quality are the important design considerations. The amplifier saturation and fiber nonlinear effects also play an important role in determining the ultimate performance of both the WDM and SCM systems. In the following subsections, we discuss the system design as well as performance degradation issues of WDM and SCM systems.

8.5.1 WDM Systems

First, the most important element in the design of WDM systems is the channel spacing, which is mainly governed by the level of interchannel crosstalk. The crosstalk leads to transfer of power from one channel to another. Crosstalk can occur as the result of many factors: (1) interchannel interference; (2) improper design of optical filters or multiplexer components; (3) amplifier-induced crosstalk; and (4) fiber nonlinearities. The amplifier-induced crosstalk and the fiber nonlinearities [54–57] were discussed in the previous chapters.

In this section, we discuss the first two factors. The first factor, the crosstalk between the signal channels due to interchannel interference or image band signal, is the dominant factor in determining the channel spacing. This problem has been studied both theoretically and experimentally in coherent WDM systems in recent years [58–61]. This crosstalk can be minimized by using the image rejection optical balanced heterodyne receivers [61]. Figure 8.12 shows the channel spacing or the number of channels in a 10 nm tuning range of the local oscillator as a function of the bit rate for various modulation formats. From Figure 8.12, it can be seen that the MSK system requires the minimum channel spacing as compared to other formats. This is because, the power spectral density of the MSK signal decays with f^{-4} and is

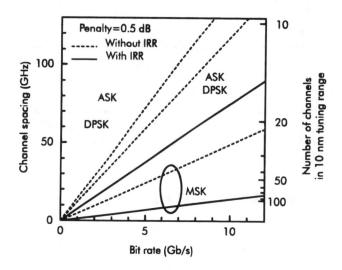

Figure 8.12 Channel spacing or the number of channels versus bit rate for various modulation formats, at the crosstalk penalty of 0.5 dB with and without image rejection receivers. (*Source:* [61]. © 1990 IEEE. Reprinted with permission.)

more compact than that of ASK or DPSK signaling schemes, which decay with f^{-2}. Additionally, the channel spacing in the MSK scheme can be narrowed to less than 1/3 when the image rejection ratio (IRR) is used, which accommodates more channels in the 10 nm tuning range.

In the case of ASK and DPSK schemes, same channel spacing is required due to the fact that the power spectral densities of ASK and PSK signals are essentially the same. On the other hand, the crosstalk penalty without the IRR is strongly dependent on the IF frequency. The theoretical results of the channel spacing with and without IRR are summarized in Table 8.4 [61]. In Table 8.4, the intermediate frequency f_{If}, B_{IF}, and the channel spacing are normalized with respect to the bit rate B. The above results assume negligible linewidth of the laser sources. However, the impact of laser phase noise on the channel spacing of coherent lightwave systems

Table 8.4
Channel Spacing With and Without IRR

Modulation format	f_{If}	B_{IF}	Spacing without IRR	Spacing with IRR
ASK	4	2	13.8	7.5
DPSK	2	2	11.5	7.5
MSK	1.75	1.5	4.84	1.43

has been carried out in [62–64] for a wide variety of IF filters, and the analysis reveals that the required channel spacing increases with the laser linewidth. For example, in the case of amplified ASK systems, for two adjacent channels with 27% linewidth of each channel, the required separation is about 24 bit rates for heterodyne detection and is about 27 bit rates for direct detection for a 1-dB sensitivity penalty. Thus, the system capacity for a multichannel direct detection and a multichannel heterodyne ASK system is about the same within a given optical bandwidth.

Next, let us look at the crosstalk induced by the multiplexer and FP filter components [65,66]. The crosstalk analysis and filter optimization of the FP filter structures was carried out in [66] in terms of the crosstalk penalty and BER. The analysis showed that the double cavity FP filters outperformed the single cavity filters, with the optimized vernier double-cavity filter producing the best performance. In the case of multiplexer design, the crosstalk penalty mainly depends on the internal geometry of the devices and the device geometry should be optimized for negligible crosstalk penalty.

Second, frequency stabilization is a design requirement for the stability of the WDM system [7]. The frequency stabilization can be achieved by using atomic or molecular transitions in several materials such as HeNe, Nd:YAG or krypton, or acetylene. The transition can be used for frequency locking of semiconductor lasers or for the development of stable lasers serving as frequency standards.

8.5.2 SCM Systems

The first design requirement for analog SCM lightwave systems is the required SNR at the receiver input, so that the replica of the transmitted analog waveform can be recovered. This requires a light source having linear PI characteristics and a linear fiber optic channel. In practice, however, the signal becomes distorted due to the system nonlinearities, causing intermodulation distortion (IMD). The severity of the IMD depends on the interchannel interference created by the second-order and third-order intermodulation products. Additionally, several other mechanisms, such as frequency chirp and fiber dispersion, can contribute to IMD.

Another important design requirement for analog SCM systems is RIN. A low RIN of less than -150 dB/Hz is usually required for a negligible power penalty. To illustrate the nonlinear power-current characteristics of the light source, the channel capacity of the system versus received power of the system is shown in Figure 8.13 as a function of fundamental and practical CNR limits. The fundamental limit in the above figure corresponds to the shot noise limit and the practical limit corresponds to the performance limit imposed by the laser RIN and the receiver thermal noise. As can be seen from the curves, the channel capacity is linearly dependent on the received power and in the case of practical limit, the channel capacity shows a nonlinear dependence on the received power. Thus, in practice, both the received power and required CNR determines the system capacity.

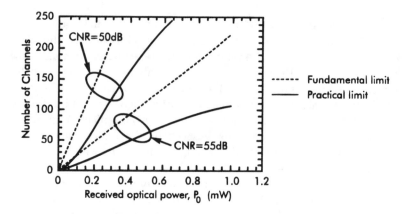

Figure 8.13 Number of channels versus the received power for fundamental and practical limits with CNR of 50 and 55 dB. (*Source:* [67]. © 1992 IEEE. Reprinted with permission.)

Another problem involved with the analog or digital SCM systems is the clipping distortion [68–70]. The clipping distortion is the result of signal waveform clipping, when the multiplexed signal level falls below the laser diode threshold. This clipping distortion at the laser transmitter imposes a fundamental limit on the channel capacity of the system. To reduce the clipping induced degradation, the multiplexed analog signal can be preclipped in order to prevent the multiplexed signal from falling below the threshold [70]. Additionally equalization technique (pretransmission compression and postdetection decompression) can be utilized to minimize the clipping distortion [69].

In the case of coherent analog SCM systems, in addition to RIN and IMD, laser phase noise plays a key role in determining the performance of the system. The coherent AM links can be designed to be insensitive to phase noise, but the coherent FM and PM links are intrinsically sensitive to phase noise because their signal information is contained in the phase noise [71]. In coherent digital SCM links, FSK transmission is commonly used [45]. In this case, the CNR is degraded by both interchannel crosstalk and IMD. The crosstalk is negligible when the channel spacing is chosen to be twice the bit rate, which is considerably smaller than that of WDM systems. The effect of IMD on the system performance is measured by the power penalty. The power penalty can be kept below 2 dB by choosing the optimum value of the phase modulation index for each channel.

In the case of SCM multi-access systems, the limiting factor is the optical beat interference due to multiple optical carriers. The experimental results have shown that the optical interference can seriously degrade the performance of an SCM system and under the worst case conditions, not even a single video channel can be transmitted [72]. The optical interference can be minimized by separating the laser wavelengths.

Other degradation factors include fiber nonlinear effects [73,74], mode partition noise and laser chirp [75], and polarization dependent distortion [76]. Fiber nonlinearities such as SRS and SBS can affect the system performance and hence limit the channel capacity of the system. The effects of SRS can be reduced substantially by allocating higher microwave subcarrier frequencies [73]. Similarly, the effects of SBS can be minimized by broadening the optical spectrum above the SBS linewidth or by distributing the optical power over a number of optical subcarriers [74]. The mode partition noise and laser chirp can be avoided by using the single longitudinal mode lasers and external modulators, respectively. Also, the polarization distortion caused by the laser chirp can be minimized by using external modulators.

References

[1] Cheung, N. K., K. Nosu, and G. Winzer, Eds., "Special issue on dense wavelength division multiplexing techniques for high capacity and multiple access communication systems," *IEEE Journal on selected areas in communications*, Vol. 8, No. 6, 1990.

[2] Darcie, T. E., K. Nawata, and J. B. Glabb, Eds., "Special issue on broad-band lightwave video transmission," *IEEE Journal of Lightwave Technology*, Vol. 11, No. 1, 1993.

[3] Karol, M. J., G. Hill, C. Lin, and K. Nosu, Eds., "Special issue on broadband optical networks," *IEEE Journal of Lightwave Technology*, Vol. 11, No. 5/6, 1993.

[4] Tucker, R. S., G. Eisenstein, and S. K. Korotky, "Optical time division multiplexing for very high bit rate transmission," *IEEE Journal of Lightwave Technology*, Vol. 6, No. 11, 1988, pp. 1737–1749.

[5] Marhic, M., "Coherent optical CDMA networks," *IEEE Journal of Lightwave Technology*, Vol. 11, No. 5/6, 1993, pp. 854–864.

[6] Brackett, C. A., "Dense wavelength division multiplexing networks: principles and applications," *IEEE Selected Areas on Communications*, Vol. 8, No. 6, 1990, pp. 948–964.

[7] Kazovsky, L. G., C. Barry, M. Hickey, C. A. Noronha, and P. Poggiolini, "WDM local area networks," *IEEE Magazine of Lightwave Telecommunication Systems*, Vol. 3, No. 2, 1992, pp. 8–15.

[8] Goodman, M. S., H. Kobrinski, M. P. Vecchi, R. M. Bulley, and J. M. Gimlett, "The LAMDANET multiwavelength network: architecture, applications, and demonstrations," *IEEE Selected Areas on Communications*, Vol. 8, No. 6, 1990, pp. 995–1004.

[9] Kazovsky, L. G., C. Barry, M. Hickey, C. A. Noronha, and P. Poggiollini, "STARNET multigigabit-per-second optical LAN utilizing passive WDM star," *IEEE Journal of Lightwave Technology*, Vol. 11, No. 5/6, 1993, pp. 1009–1027.

[10] Arthurs, E., M. S. Goodman, H. Kobrinski, M. Tur, and M. P. Veechi, "Multiwavelength optical crossconnect for parallel processing computers," *Electronics Letters*, Vol. 24, 1986, pp. 119–120.

[11] Arthurs, E., M. S. Goodman, H. Kobrinski, and M. P. Veechi, "HYPASS: An optoelectronic hybrid packet-switching system" *IEEE Selected Areas on Communications*, Vol. 6, 1988, pp. 1500–1510.

[12] Wagner, S. S., H. Kobrinski, T. J. Robe, H. L. Lemberg, and L. S. Smoot, "A passive photonic loop architecture employing wavelength-division multiplexing," in *Conf. Proc., GLOBECOM '88*, 1988, pp. 1569–1573.

[13] Gidon, R., et al., "TeraNet: A multi gigabit per second hybrid circuit/packet switched lightwave network," *Proc. SPIE Adv. Fiber Commun. Tech.*, Boston, 91, pp. 40–48.

[14] Dono, N. R., et al., "A wavelength division multiple access network for computer communications," *IEEE Selected Areas on Communications*, Vol. 8, No. 6, 1990, pp. 983–984., pp. 1500–1510.

[15] Fioretti, A., et al., "An evolutionary configuration for an optical coherent multichannel network," *GLOBECOM '90*, San Diego, 1990, pp. 779–783.

[16] Toba, H., et al., "A 100 channel optical FDM transmission/distribution at 622 Mb/s over 50 km," *IEEE Journal of Lightwave Technology*, Vol. 8, No. 9, 1990, pp. 1396–1401.

[17] Wagner, R. E., et al., "16 channel coherent broadcast network at 155 Mb/s," *OFC '89*, Houston, Paper PD12, 1989.

[18] Mollenauer, L. F., S. G. Evangelides, and J. P. Gordon, "Wavelength division multiplexing with solitons in ultra-long distance transmission using lumped amplifier," *IEEE Journal of Lightwave Technology*, Vol. 9, No. 3, 1991, pp. 362–367.

[19] Moores, J. D., "Ultra-long distance wavelength-division-multiplexed soliton transmission using inhomogeneously broadened fiber amplifiers," *IEEE Journal of Lightwave Technology*, Vol. 10, No. 4, 1992, pp. 482–487.

[20] Mollenauer, L. F., E. Lichtman, G. T. Harvey, M. J. Neubelt, and B. M. Nyman, "Demonstration of error-free soliton transmission over more than 15000 km at 5 Gb/s single channel and over more than 1000 km at 10 Gb/s in two channel WDM," *Electronics Letters*, Vol. 28, No. 8, 1992, pp. 792–794.

[21] Kaminow, I. P., "FSK with direct detection in optical multiple-access FDM networks," *IEEE Selected Areas on Communications*, Vol. 8, No. 6, 1990, pp. 1005–1014.

[22] Kolltveit, E., B. Biotteau, I. Riant, et al., "Soliton frequency-guiding by UV-written fiber Fabry-Perot filter in a 2 × 5 Gb/s wavelength division multiplexing transmission over transoceanic distances," *IEEE Photonics Technology Letters*, Vol. 7, No. 12, 1995, pp. 1500.

[23] Heismann F., et al., "Narrow-linewidth electro-optically tunable InGaAsP Ti:linbo extended cavity laser," *Applied Physics Letters*, Vol. 51, 1987, pp. 164–165.

[24] Okai, M., S. Sakano, and N. Chinone, "Wide-range continuous tunable double-sectioned distributed feedback lasers," *15th European Conference Optic. Commun.*, Sweden, 1989.

[25] Alferness, R. C., "Widely tunable InGaAsP/InP laser based on a vertical coupler intractivity filter," *OFC '92*, Postdeadline Papers, Paper PD2, San Jose, 1992, pp. 321–324.

[26] Imai, H., "Tuning results of 3-sectioned DFB lasers," *Semiconductor laser workshop, CLEO '89*, Baltimore, 1989.

[27] Lee, T. P. and C. E. Zah, "Wavelength tunable and single frequency semiconductor lasers for photonic communications networks," *IEEE Communications Magazine*, 1989, pp. 42–52.

[28] Koch, L., et al., "Continuously tunable 1.5 μm multiple quantum well GaInAs/GaInAsP distributed-bragg-reflector lasers," *Electronics Letters*, Vol. 24, No. 23, 1988.

[29] Kobayashi, K. and I. Mito, "Single frequency and tunable laser diodes," *IEEE Journal of Lightwave Technology*, Vol. LT-6, No. 11, 1988, pp. 1623–1633.

[30] Chung, Y. C., et al., "WDM coherent star network with absolute frequency reference," *Electronics Letters*, Vol. 24, No. 21, 1988, pp. 1313–1314.

[31] Sudo, S., Y. Saki, and T. Ikegami, "Frequency stabilization for DFB lasers diodes using acetylene absorption lines in 1.51–1.55 μm," *OFC '91*, San Diego, 1991, Paper FB7.

[32] Kobrinski, H. and K. W. Cheung, "Wavelength-tunable optical filters: applications and technologies," *IEEE Communications Magazine*, 1989, pp. 53–63.

[33] Numai, T., S. Muratam, T. Sasaki, and I. Mito, "1.5 μm tunable wavelength filter using phase shift controllable DFB LD with wide tuning range and high constant gain," in *Conf. Proc. ECOC, '88*, UK, pp. 243–246.

[34] Takato, N., et al., "Silica-based integrated optic Mach-Zehnder multi/demultiplexer family with channel spacing of 0.01–250 nm," *IEEE J. on Selected Areas in Communications*, Vol. 8, No. 6, 1990, pp. 1120–1127.

[35] Olshansky, R., V. A. Lanzisera, and P. A. Hill, "Subcarrier multiplexed systems for broad-band distribution," *IEEE Journal of Lightwave Technology*, Vol. 7, No. 9, 1989, pp. 1329–1341.

[36] Olshansky, R. and E. Eichen, "Microwave-multiplexed wideband lightwave systems using optical amplifiers for subscribe distribution," *Electronics Letters*, Vol. 24, 1988, pp. 922–923.

[37] Darcie, T. E., "Subcarrier multiplexing for multiple access lightwave networks," *IEEE Journal of Lightwave Technology*, Vol. LT-5, 1987, pp. 1103–1110.

[38] Way, W. I., "Subcarrier multiplexed lightwave system design considerations for subscriber loop applications," *IEEE Journal of Lightwave Technology*, Vol. 7, 1989, pp. 1806–1818.

[39] Olshansky, R., V. A. Lanzisera, S. F. Su, R. Gross, et al., "Subcarrier multiplexed road-band service network: A flexible platform for broad-band subscriber services," *IEEE Journal of Lightwave Technology*, Vol. 11, 1993, pp. 60–69.

[40] Olshansky, R. and V. Lanzisera, "60 channel FM video subcarrier multiplexed optical communication system," *Electronics Letters*, Vol. 23, 1987, pp. 1196–1197.

[41] Gabla, P. M., et al., "35 AM-VSB TV channels distribution with high signal quality using a 1480 nm diode pumped fiber postamplifier," *IEEE Photonics Technology Letters*, Vol. 3, 1991, pp. 56–58.

[42] Way, W. I., "160 channel FM-video transmission using optical FM/FDM subcarrier multiplexing and an erbium doped optical fiber amplifier," *Electronics Letters*, Vol. 26, 1990, pp. 139–142.

[43] Yoneda, E., K. Suto, K. Kikushima, and H. Yoshinaga, "All-fiber video distribution systems using SCM and EDFA techniques," *IEEE Journal of Lightwave Technology*, Vol. 11, 1993, pp. 128–137.

[44] Hill, P. and R. Olshansky, "Twenty channel FSK subcarrier multiplexed optical communication system for video distribution," *Electronics Letters*, Vol. 24, 1988, pp. 892–893.

[45] Gross, R., and R. Olshansky, "Multichannel coherent FSK experiments using subcarrier multiplexing techniques," *IEEE Journal of Lightwave Technology*, Vol. 8, 1990, p. 406.

[46] Hill, P., and R. Olshansky, "8 Gb/s subcarrier multiplexed coherent lightwave system," *IEEE Photonics Technology Letters*, Vol. 3, 1991, pp. 764–766.

[47] Gross, R. and R. Olshansky, "Optical DQPSK video system with heterodyne detection," *IEEE Photonics Technology Letters*, Vol. 3, 1991, pp. 262–264.

[48] Watanabe, S. et al., "Optical coherent broad-band transmission for long-haul distribution systems using subcarrier multiplexing," *IEEE Journal of Lightwave Technology*, Vol. 11, No. 1, 1993, pp. 116–127.

[49] Olshansky, R. and V. Lanzisera, "Subcarrier multiplexed passive optical network for low-cost video distribution," *Optical Fiber Communication Conference*, Houston, 1989.

[50] Olshansky, R., and E. Eichen, "Microwave-multiplexed wide-band lightwave system using optical amplifiers for subscribe distribution," *Electronics Letters*, Vol. 24, 1988, p. 922.

[51] Wood, T. H., et al., "Demonstration of a cost effective, broadband passive optical network system," *IEEE Photonics Technology Letters*, Vol. 6, 1994, pp. 575–577.

[52] Feldman, R. D., T. H. Wood, and R. F. Austin, "Operation of a frequency shift keyed subcarrier multiple-access system for a passive optical network in the presence of strong adjacent channel interference," *IEEE Photonics Technology Letters*, Vol. 7, 1995, pp. 427–430.

[53] Wagner, S. S. and R. C. Menendez, "Evolutionary architectures and techniques for video distribution on fiber," *IEEE Commun Mag.*, Vol. 27, No. 2, 1989, pp. 17–25.

[54] Shibata, N., K. Nosu, K. Iwashita, and Y. Azuma, "Transmission limitations due to fiber nonlinearities in optical FDM systems," *IEEE Selected Areas on Communications*, Vol. 8, No. 6, 1990, pp. 1068–1077.

[55] Tomita, A. "Crosstalk caused by stimulated Raman scattering in single-mode wavelength division multiplexed systems," *Optics Letters*, Vol. 8, No. 7, 1983, pp. 412–414.

[56] Maeda, et al., "The effect of four-wave mixing in fibers on optical frequency division multiplexed systems," *IEEE Journal of Lightwave Technology*, Vol. 8, 1990, pp. 1402–1408.

[57] Inoue, K. and H. Toba, "Error rate degradation due to fiber four-wave mixing in 4-channel FSK direct detection transmission," *IEEE Photonics Technology Letters*, Vol. 3, 1991, pp. 77–79.

[58] Suyama, M., T. Chikama, and H. Kuwahara, "Channel allocation and crosstalk penalty in coherent optical frequency division multiplexing systems," *Electronics Letters*, Vol. 24, No. 20, 1988, pp. 1278–1279.

[59] Kazovsky, L. G., and J. L. Gimlett, "Sensitivity penalty in multichannel coherent optical communications," *IEEE Journal of Lightwave Technology*, Vol. 6, No. 9, 1988, pp. 1353–1365.

[60] Toba, H. et al., "Factors affecting the design of optical FDM information distribution systems," *IEEE Selected Areas on Communications*, Vol. 8, No. 6, 1990, pp. 965–972.

[61] Chikama, T., et al., "Optical heterodyne image-rejection receiver for high density optical frequency division multiplexing systems," *IEEE Selected Areas on Communications*, Vol. 8, No. 6, 1990, pp. 1087–1094.

[62] Jacobsen, G., and I. Garrett, "The effect of crosstalk and phase noise in multichannel coherent optical ASK systems," *IEEE Journal of Lightwave Technology*, Vol. 9, No. 8, 1991, pp. 1006–1018.

[63] Jacobsen, G., and I. Garrett, "The effect of crosstalk and phase noise in multichannel coherent optical DPSK systems with tight IF filtering," *IEEE Journal of Lightwave Technology*, Vol. 9, No. 11, 1991, pp. 1609–1617.

[64] Jacobsen, G., and I. Garrett, "Crosstalk and phase noise effects in multichannel coherent optical CPFSK systems with tight IF filtering," *IEEE Journal of Lightwave Technology*, Vol. 9, No. 9, 1991, pp. 1168–1177.

[65] Hill, A. M., and D. B. Payne, "Linear crosstalk in wavelength division multiplexed optical fiber transmission systems," *IEEE Journal of Lightwave Technology*, Vol. LT-3, No. 3, 1985, pp. 643–650.

[66] Humblet, P. A. and W. M. Hamdy, "Crosstalk analysis and filter optimization of single and double cavity Fabry-Perot filters," *IEEE Selected Areas on Communications*, Vol. 8, No. 6, 1990, pp. 1087–1094.

[67] Chung, C. J. and I. Jacobs, "Practical TV channel capacity of lightwave multichannel AM SCM systems limited by the threshold nonlinearity of laser diodes," *IEEE Photonics Technology Letters*, Vol. 4, No. 3, 1992, pp. 289–292.

[68] Frigo, N. J. and G. E. Bodeep, "Clipping distortion in AM-VSB CATV subcarrier multiplexed lightwave systems," *IEEE Photonics Technology Letters*, Vol. 4, No. 7, 1992, pp. 781–784.

[69] Ho, K. P. and J. M. Kahn, "Equalization technique to reduce clipping-induced nonlinear distortion in subcarrier multiplexed lightwave systems," *IEEE Photonics Technology Letters*, Vol. 5, No. 9, 1993, pp. 1100–1103.

[70] Kanawaza, A., M. Shibutani, and K. Emura, "Pre-clipping method to reduce clipping-induced degradation in hybrid analog/digital subcarrier multiplexed optical transmission systems," *IEEE Photonics Technology Letters*, Vol. 7, No. 9, 1995, pp. 1069–1071.

[71] Kalman, R. F., J. C. Fan, and L. G. Kazovsky, "Dynamic range of coherent analog fiber optic links," *IEEE Journal of Lightwave Technology*, Vol. 12, No. 7, 1994, pp. 1263–1277.

[72] Desem, C. "Measurement of optical interference due to multiple optical carriers in subcarrier multiplexing," *IEEE Photonics Technology Letters*, Vol. 3, No. 4, 1991, pp. 387–389.

[73] Wang, Z. et al., "Performance limitations imposed by stimulated Raman scattering in optical WDM SCM video distribution systems," *IEEE Photonics Technology Letters*, Vol. 7, No. 12, 1995, pp. 1492–1494.

[74] Wiullems, F. W. et al., "Simultaneous suppression of stimulated Brillouin scattering and interferometrioc noise in externally modulated lightwave AM-SCM systems," *IEEE Photonics Technology Letters*, Vol. 6, No. 12, 1994, pp. 1476–1478.

[75] Kikushima, K., K. Suto, H. Yoshinaga, and E. Yoneda, "Polarization dependent distortion in AM-SCM video transmission systems," *IEEE Journal of Lightwave Technology*, Vol. 12, No. 4, 1994, pp. 650–657.

[76] Meslener, G. J. "Mode partition noise in microwave subcarrier transmission systems," *IEEE Journal of Lightwave Technology*, Vol. 12, No. 1, 1994, pp. 118–126.

[77] Kashima, N., *Optical Transmission for the Subscriber Loop*, Norwood, MA: Artech House, 1993.

[78] Personick, S. D., "Towards Global Information Networking," *Proc. of IEEE*, Vol. 81, No. 11, 1993, pp. 1549–1557.

List of Useful Physical Constants

q	Electron charge	1.6×10^{-19} C
c	Velocity of light in vacuum	3.0×10^{8} m/s
h	Planck's constant	6.62×10^{-34} Js
k_B	Boltzmann constant	1.381×10^{-23} J/K
μ_0	Permeability of free space	$4\pi \times 10^{-7}$ H/m
ϵ_0	Permittivity of free space	8.85×10^{-12} F/m

About the Author

Rajappa Papannareddy was born in Bangalore, India. He received the B.E. degree in electronics from Bangalore University in 1975; the M.S.E.E. from the University of Maryland, College Park in 1983; and the Ph.D degree in electrical engineering from Southern Methodist University in 1987. His Ph.D. thesis was conducted in the area of semiconductor lasers under the guidance of Professor Jerome K. Butler.

Dr. Papannareddy worked as an engineer from 1976 to 1984 at Indian Space Research Organization Satellite Center, Bangalore, India. He was involved with the development and test and evaluation of satellite telemetry, tracking, and command subsystems. Additionally, he was involved with reliability analysis and the electromagnetic compatibility design of satellite systems.

Currently, Dr. Papannareddy is working as an associate professor at Purdue University (North Central campus). He teaches courses in the areas of circuit theory, analog electronics, and communication systems. His research interests include the area of lightwave communication systems relating to the analysis of semiconductor lasers, optical amplifiers, and direct and coherent detection schemes. He has published several journal papers and presented many papers in technical conferences. His research work has been supported by summer faculty grants from Purdue Research Foundation. He has received a National Science Foundation award to set up the microcomputer-based electronic instrumentation laboratory. He is a member of the IEEE and ASEE professional organizations. He has reviewed papers for several publications, including the *IEEE Journal of Quantum Electronics*.

Index

The Artech House Optoelectronics Library

Brian Culshaw, Alan Rogers, and Henry Taylor, *Series Editors*

Acousto-Optic Signal Processing: Fundamentals and Applications, Pankaj Das

Amorphous and Microcrystalline Semiconductor Devices, Volume II: Materials and Device Physics, Jerzy Kanicki, editor

Bistabilities and Nonlinearities in Laser Diodes, Hitoshi Kawaguchi

Chemical and Biochemical Sensing With Optical Fibers and Waveguides, Gilbert Boisdé and Alan Harmer

Coherent and Nonlinear Lightwave Communications, Milorad Cvijetic

Coherent Lightwave Communication Systems, Shiro Ryu

Elliptical Fiber Waveguides, R. B. Dyott

Field Theory of Acousto-Optic Signal Processing Devices, Craig Scott

Frequency Stabilization of Semiconductor Laser Diodes, Tetsuhiko Ikegami, Shoichi Sudo, Yoshihisa Sakai

Fundamentals of Multiaccess Optical Fiber Networks, Denis J. G. Mestdagh

Germanate Glasses: Structure, Spectroscopy, and Properties, Alfred Margaryan and Michael A. Piliavin

High-Power Optically Activated Solid-State Switches, Arye Rosen and Fred Zutavern, editors

Highly Coherent Semiconductor Lasers, Motoichi Ohtsu

Iddq Testing for CMOS VLSI, Rochit Rajsuman

Integrated Optics: Design and Modeling, Reinhard März

Introduction to Lightwave Communication Systems, Rajappa Papannareddy

Introduction to Glass Integrated Optics, S. Iraj Najafi

Introduction to Radiometry and Photometry, William Ross McCluney

Introduction to Semiconductor Integrated Optics, Hans P. Zappe

Laser Communications in Space, Stephen G. Lambert and William L. Casey

Optical Fiber Amplifiers: Design and System Applications, Anders Bjarklev

Optical Fiber Communication Systems, Leonid Kazovsky, Sergio Benedetto, Alan Willner.

Optical Fiber Sensors, Volume Two: Systems and Applicatons, John Dakin and Brian Culshaw, editors

Optical Fiber Sensors, Volume Three: Components and Subsystems, John Dakin and Brian Culshaw, editors

Optical Interconnection: Foundations and Applications, Christopher Tocci and H. John Caulfield

Optical Network Theory, Yitzhak Weissman

Optoelectronic Techniques for Microwave and Millimeter-Wave Engineering, William M. Robertson

Reliability and Degradation of LEDs and Semiconductor Lasers, Mitsuo Fukuda

Reliability and Degradation of III-V Optical Devices, Osamu Ueda

Semiconductor Raman Laser, Ken Suto and Jun-ichi Nishizawa

Semiconductors for Solar Cells, Hans Joachim Möller

Smart Structures and Materials, Brian Culshaw

Ultrafast Diode Lasers: Fundamentals and Applications, Peter Vasil'ev

For further information on these and other Artech House titles, contact:

Artech House
685 Canton Street
Norwood, MA 02062
617-769-9750
Fax: 617-769-6334
Telex: 951-659
email: artech@artech-house.com

Artech House
Portland House, Stag Place
London SW1E 5XA England
+44 (0) 171-973-8077
Fax: +44 (0) 171-630-0166
Telex: 951-659
email: artech-uk@artech-house.com

WWW: http://www.artech-house.com